T0331425

The Deuteromycetes

Mitosporic Fungi

Classification and Generic Keys

The Deuteromycetes
Mitosporic Fungi
Classification and Generic Keys

E. Kiffer
Unitè associèe de Biologie forestière
Universitè Nancy I
Facultè des Sciences
Boulevard des Aiguillettes
B.P. 239
54506, VANDOEUVRE-lès-NANCY
Cedex, FRANCE

M. Morelet
Unitè de recherches Ecosystèmes forestiers
INRA (Institut National de la Recherche Agonomique)
54280 CHAMPENOUX
FRANCE

Foreword by
Professor G.L. Hennebert

CRC Press
Taylor & Francis Group
Boca Raton   London   New York

CRC Press is an imprint of the
Taylor & Francis Group, an informa business

Science Publishers Inc.
U.S.A.

Reprinted 2009 by CRC Press

CRC Press
6000 Broken Sound Parkway, NW
Suite 300, Boca Raton, FL 33487
270 Madison Avenue
New York, NY 10016
2 Park Square, Milton Park
Abingdon, Oxon OX14 4RN, UK

SCIENCE PUBLISHERS, INC.
Post Office Box 699
Enfield, New Hampshire 03748
United States of America

ISBN 1-57808-068-1

Library of Congress Cataloging-in-Publication Data

Kiffer, E.
    [Deutromycètes. English]
    The Deuteromycetes, mitosporic fungi: classification
and generic keys/E. Kiffer, M. Morelet; foreword by
G.L. Hennebert.
    p.cm.
    Includes bibliographical references (p.) and index.
    ISBN 1-57808-068-1
    1. Fungi imperfecti—Identification. 2. Fungi im-
perfecti—Classification. I. Morelet, Michel.
II. Title.

QK625.Al K5413 1999
579.5'5—dc21
                                      99-045523

Aidé par le ministère français chargé de la culture.
Published with the support of the French Ministry of Culture.

Translation of : **les deutéromycètes** classification et clés d'identification
            générique, INRA, Paris, 1997. Text updated by the
            authors for the English edition in 1999.

French edition : © INRA, Paris, 1997
                  ISBN 2-7380-0729-5
                  ISSN 1150-3564

Translation Editor : Latha Anantharaman
Technical Editor   : Dr. H.C. Dube

For many people, the word 'fungus' suggests the fructifications of various forms and colours found chiefly in the autumn in the forests and nearby areas. These are the carpophores, essentially of Basidiomycetes (e.g., *Amanita*, boletus, chanterelle) and sometimes of Ascomycetes (e.g., morels, truffles).

Apart from these commonly seen forms, there are many other fungi that are often microscopic.

The Deuteromycetes are among these and represent the phase of asexual multiplication (anamorph) of the higher fungi (mostly Ascomycetes, and to a lesser extent Basidiomycetes).

They represent the second group of fungi in numerical significance (around 2400 genera and 20,000 species) after the Ascomycetes.

Whether one studies the fungal flora of the air, litter, soil, or any other substrate (faeces, diseased plants), it is these, primarily, that are observed, *in situ* or isolated on nutritive media.

From this omnipresence arises also their **economic importance**. We can cite, for example, the case of the genus *Acremonium*, of which certain species live in the stems of Gramineae as endophytes toxinogenic to cattle. These toxicoses lead to losses (lower production of meat and milk, even abortion) estimated at 760 million dollars a year in the United States.

The work we present here is based on the modern classification of this group. Barring some tropical examples, it covers essentially genera found in the temperate zone of the Northern hemisphere, and we hope it will help English-reading users to identify them.

FOREWORD

The world of fungi, which includes moulds and yeasts, is vast. Some 76,000 species are known, but it is estimated to comprise 1,500,000 species.

This world is understood, by the nature of its characters, to be a living world apart, a kingdom apart among the living kingdoms, a kingdom that is not plant or animal, but fungal. One of the important characters that distinguishes fungi from plants is the composition of their cell walls, made for the most part of chitin, of a type similar to that of Arthropoda. Being heterotrophic, like animals, the fungi have, like animals, four modes of life: parasitism on other living beings (animals, plants, fungi, or protists), symbiosis with plants and animals, association with other living beings that can become predatory, and saprophytism, or life that is dependent on the degradation of dead organic matter.

Our knowledge of fungal diversity is based on observation, taxonomy, nomenclature, classification, and identification.

Taxonomy is the characterization and distinction of taxa (species) based on the observation of precise characters. It is the recognition of individuals that resemble each other to the point of forming a homogeneous group having certain common characters and differing from other groups of individuals on the basis of other characters. One individual is chosen in this homogeneous group (taxon) as a specimen of reference, the type. A proper name is given to this type individual representing the taxon and constitutes an essential part of the name of the species.

Classification is the placing of species in a hierarchical system of morphological and biological resemblance. The species is classed in a genus and is given the name of the genus. The species name along with the genus name forms the binomial, the name that designates the species.

Nomenclature regulates the constitution of names, their validity, their legitimacy, and their priority or synonymy, and maintains a single correct name for each taxon.

Identification is the decision to consider an individual as belonging to a known taxon and attributing to it the name of this taxon. If no identification can be made, the specimen can represent a taxon not yet described, a new species.

Taxonomy, classification, nomenclature, and identification therefore depend on the observation of the characters of individuals. These characters are morphological and biological. According to Linnaeus, the classification of plants is based on the morphology of their organs of sexual reproduction.

Among the fungi, as among other cryptogams, sexual reproduction was long unknown and is still unknown in a good number. The taxonomy and classification are therefore based, in a Linnaean system, on all other forms of reproduction. The great diversity of forms and reproductive organs in the fungi thus form the origin of the distinction of a large number of species. After the first discoveries of sexuality in the fungi around the 1860s, the forms of reproduction could be recognized as asexual or sexual. The asexual species were grouped by Fuckel in 1869 in the category of *Fungi Imperfecti*, imperfect fungi (*unvollständiger Pilze*) and the sexual species in the Fungi Perfecti, perfect fungi (*vollkommene Pilze*). Saccardo picked up the name *Fungi Imperfecti* around 1879 and modified it to '*Imperfectae* Fuckel' or *Fungi Inferiores*. In 1899, Saccardo introduced the denomination *Deuteromycetae*, for the *Fungi secondarii vel inferiores* (deuteros = secondary), otherwise called the *Fungi Imperfecti*.

Apart from the Deuteromycetes, Saccardo classed the fungi without any known mode of reproduction other than surviving mycelial organs in a category that he called *Mycelia sterilia* in 1881. This group was later called Agonomycetales and considered part of the Deuteromycetes.

The brothers Louis-René and Charles Tulasne in 1851 to 1865 and Anton de Bary in 1866 to 1884 demonstrated that the two types of reproduction, sexual and asexual, could coexist, simultaneously or successively, in many fungi. It was then realized that the names of asexual species were only names given to the secondary forms of reproduction (imperfect forms) belonging to the fungi otherwise given a name typified by sexual reproduction (perfect form). There arose thus two parallel nomenclatures for the fungi that were simultaneously sexual and asexual and the fungi that were sexual or asexual.

This double nomenclature was accepted by the International Code of Botanical Nomenclature around 1906, as a tolerated exception to its principle of 'one and only one name for an organism'. The tolerance was motivated by two reasons. One was the necessity of classifying in a coherent manner all the fungi known only in the asexual form (*Fungi Imperfecti*), which cannot be classified among the higher fungi (called *Perfecti*), and, in order to help in their diagnosis, of classifying them together with the asexual forms already named among the higher fungi. The other reason, which is practical, was to ensure a classification enabling the identification of any fungus, perfect or imperfect, that presents, at the moment of its observation, only an asexual reproduction. This tolerance of a double nomenclature is currently regulated by article 59 of the Code of Botanical Nomenclature.

In order to avoid this anthropomorphism of designating certain fungi as imperfect because they do not manifest sexuality in their spore production or, to be honest, because they are imperfectly known, the terms 'anamorph', 'holomorph', and 'teleomorph' were introduced in 1977. Anamorph to designate the secondary form (*ana* = near) of a fungus denominated Deuteromycete for its asexual or imperfect reproduction; teleomorph for the form of sexual reproduction of the same fungus; and holomorph to designate the fungus denominated Ascomycete or Basidiomycete in its entirety. As early as 1863, the Tulasne brothers distinguished different names given to the same fungus in names designating 'the imperfect fungus, conidial' on the one hand, those designating 'the perfect fungus, ascophore' on the other, and the name of the 'entire fungus'.

Since then, more and more species of Ascomycetes and Deuteromycetes have been recognized as being organically related and thus constituting a single species (holomorphic) in place of two or three, and therefore having only one name. These organic connections of forms of reproduction are now reinforced with genomic identity of the teleomorph and of a corresponding anamorph or anamorphs.

Also, the current trend is to analyse the technical possibilities of integrating the two parallel fungal taxonomies, those of sexual fungi (Ascomycetes and Basidiomycetes) and those of asexual fungi and the anamorphic forms of sexual fungi (Deuteromycetes), into a single taxonomy and a single nomenclature.

Nevertheless, this prospect of integration does not detract from the intrinsic and practical value of the taxonomy of Deuteromycetes based on a characterization of asexual forms of reproduction. On the contrary, whether or not this taxonomic and nomenclatural integration is achieved, the characterization of asexual forms and the contribution of taxonomic criteria of Deuteromycetes will be useful for taxonomy and the classification of sexual forms of Ascomycetes and Basidiomycetes.

It is also important to develop works such as these, which, from a viewpoint that is more extensive and deeper than possible in the morphological characterization of forms of asexual reproduction of fungi, can contribute to the taxonomy even of higher fungi.

The Deuteromycetes comprise 1700 genera of Hyphomycetes and 700 genera of Coelomycetes that cover some 20,000 known species. Their identification, traditionally done by means of characters drawn from the morphology of fruit bodies and conidia, is much refined by the observation of modes of conidiogenesis.

The morphological characters of fruit bodies and conidia have been used for ages and were well exploited by Saccardo. But they are not enough. As early as 1910, Vuillemin sought to refine the taxonomy of

Saccardo by the observation and characterization of conidiogenesis, in which he recognized several modes. In 1953, Hughes described, in a coherent system of classification of Hyphomycetes, eight generic modes of conidiogenesis, some with several modalities. Tubaki, in 1958, added a ninth mode. Since then, as intermediate modes of conidiogenesis have been discovered, this system of characterization remains of major importance. It was used by Sutton for the Coelomycetes and by Samson for the yeasts (Blastomycetes).

As has been shown by recent authors, the system can be refined further. This necessitates a detailed study of successive events that make up conidiogenesis and the establishment of an appropriate terminology for the description of each of these particular and united events. But that is the domain of research.

This work, which integrates the progress of research, is a tool for the identification of the diverse world of microfungi. It is also an excellent technical manual that proposes a unified taxonomic system of the Deuteromycete group, integrating Hyphomycetes and Coelomycetes, and based on the morphology of the conidial apparatus and of conidiogenesis with a view to their generic identification. This system also reflects the personal views of the authors on the phenomena of conidiogenesis and thereby has some originality in comparison with the interpretations of other authors.

The conidiogenetic system proposed also has certain peculiarities. It distinguishes the two basic modes of conidiogenesis, the blastic and the arthric, and recognizes an intermediate mode presently denominated 'arthroblastic' as 'meristem arthrospores'. By this the authors extend the concept of the meristem arthrospore mode of Hughes, producing a progressive basipetal chain of arthroblastic conidia, to a retrogressive conidiogenesis from a preformed conidiogenous hypha. It is true in fact that in retrogressive conidiogenesis, a certain progressive growth is observed on account of the swelling of the future conidium.

Another peculiarity is the grouping of the blastosporous fungi producing blastic conidia in acropetal chains under the denomination of Acroblastoporae (Blastosporae *sensu stricto*), as did Hughes, without distinguishing those that produce several, ramified chains by a sympodial process from those that produce a single and non-ramified chain.

Finally, another characteristic of the system is the distinction between Annellosporae and Annelloblastosporae (holoblastic) and Annellidae (enteroblastic), the former producing a succession of holoblastic conidia on a conidiogenous cell each time renewed by percurrent proliferation, and the latter producing enteroblastic conidia by renewal of the single tip of the same conidiogenous cell. It is true, according to the classified genera, and even in the authors' opinion, that it seems difficult in certain cases to distinguish something that is here named an annellophore from an annellide on the one hand and to recognize the conidiogenesis by percurrence when the conidia remain attached by a lateral fragment of the wall of the mother cell, giving it the appearance of a sympodula. The light microscope in fact does not always enable us to elucidate the mysterious plasticity of conidiogenesis. But this difficulty does not diminish the authors' attempt to present a refined system.

The value of this work lies also in the intégration of all the asexual fungi in a conidiogenetic system without giving a discriminatory priority to the organization of fruit body that separates the classic groups of Hyphomycetes and Coelomycetes. The system proposed puts on a second plane the organization of conidiophore hyphae in fruit bodies. The authors present various conidial apparatuses from the most simple (hyphal) to the most complex (conidiomal) in a quasi-continuous series from the micronematous conidiophore, to the macronematous conidiophore, the coremium, the sporodochium, the acervulus, the cupule, the thyriopycnidium, the pseudopycnidium, and the pycnidium. Such an option no longer really justifies the distinction of the taxonomic groups Hyphomycetes and Coelomycetes, more so because in particular conditions, in culture for example, the complexity of the fruit body may be simplified.

The recognition of the complex phenomena of conidiogenesis is certainly a great step towards a more natural classification of these secondary forms of reproduction. It is only by the presentation of increasingly extensive classification systems, such as that presented in this work, that we can progress towards a natural phylogeny of these forms of fungi and their relation to sexual fungi.

Finally, the great appeal of this work lies in the production of keys and simple diagrams for the identification of a great many genera. The authors have achieved their objective of presenting a practical and accessible tool of identification.

This work, because it is so abundantly illustrated, cannot but stimulate and encourage more minute observations and their precise representation by its design as much as, if not more than, the text. The Swede Elias Fries said as early as 1849 that 'it is greatly desirable that illustrators (of fungi) clearly demonstrate to our eyes the metamorphoses of all genera of fungi, since words cannot explain everything.' And Louis-René and Charles Tulasne, whose illustrations of these fungal metamorphoses are of an extraordinary quality, wrote in 1861 at the end of their work: 'No doubt, in order to study the hidden marvels of these fungi, one must devote a great deal of labour and patience, but in gazing upon them when one discovers them, how much greater is the joy!'

Grégoire Laurent Hennebert
Professor Emeritus of the University of Louvain

ACKNOWLEDGEMENTS

We thank Ms. Monique Jacquemin (Université Nancy I) for her critical reading and judicious advice on chapter I.

We also thank Dr. Walter Gams (CBS, Baarn) for his bibliographical information; Professor André Mourey (Université Nancy I) for his remarks on the first two chapters; Mr. Otto Reisinger (Université Nancy I) for useful discussions on the ultrastructure of conidiogenesis; Mr. Pascal Frey (INRA Nancy), who reviewed the text on molecular biology; and Ms. C.J.K. Wang (State University of New York, Syracuse, NY) for the communication of an article before publication.

Our thanks are due also to Ms. Lysiane Laviron (Université Nancy I) and Ms. Anne-Marie Meyer (INRA Nancy) for their assistance in typing and to Mr. Jean-Emmanuel Ménard (INRA Nancy) for his assistance in data processing.

CONTENTS

IMPACT AND DISTRIBUTION OF DEUTEROMYCETES

Practical role of Deuteromycetes

Biodegradation

In the natural environment, Deuteromycetes contribute, along with other organisms, to the biodegradation and recycling of organic matter (e.g., litter, wood). They are of various kinds:
• Cellulolytic (*Malbranchea, Stephanosporium, Oidiodendron, Chrysosporium...*)
• Lignolytic (*Arthrographis, Ptychogaster...*)
• Ligninolytic (*Geniculosporium, Spiniger, Sporotrichum pulverulentum...*)
The Deuteromycetes are a fascinating part of the succession of groups of organisms that cause degradation. They also play a role in the treatment of certain raw material, waste treatment, and the breaking down of pesticides.

Food and industry

Certain Deuteromycetes are important in nutrition and in industrial production.
• Some species of *Penicillium* are used in cheese production (*P. camembertii, P. roquefortii*). *Botrytis cinerea* is the 'noble rot' which leads to the production of sweet wines.
• *Aureobasidium pullulans*, present in the merrains of oak seasoned in the open air, improves the gustative qualities of wine in new barrels.

• Some species of *Aspergillus* and *Penicillium* produce organic acids (citric, gluconic) and pigments. Certain pigments are also produced by some species of *Helminthosporium* and *Fusarium*. Fungal biomass is produced from *Penicillium chrysogenum* to be used as fertilizer (Biosol), or as cattle feed. This same fungus is one component of Biosorbant M, which enables the extraction of uranium and radium from atomic industrial waste water.

Biomedicine

In the field of biomedicine, Deuteromycetes are among the producers of **antibiotics**, of which the penicillins (antibacterial agents produced by *Penicillium chrysogenum*) are among the best known examples. Among the antifungals, the best known is the griseofulvin, produced by *Penicillium griseofulvum*. Several antitumorals and antivirals of fungal origin are always under investigation.

Among the **immunoregulators**, one can cite the cyclosporines produced by various Hyphales, of which cyclosporine A is the most often used on account of its powerful immunosuppressant activity.

Forty per cent of the **enzymes** produced industrially are of fungal origin, most of them from Deuteromycetes, and they represent an economically very significant market.

Biocontrol

Since the success of **entomopathogens** (e.g., *Beauveria bassiana*), the industrial production of bio-insecticides is in full swing. Genera such as *Arthrobotrys, Dactylella, Dactylaria* and *Monacrosporium* are **predators** used against pathogenic nematodes. Among the numerous **mycoparasite** genera or fungal antagonists, some are used in biocontrol (*Trichoderma viride* and *Verticillium biguttatum* against *Rhizoctonia* in soil; *Scytalidium* against *Phellinus weirii*, which causes rot of telephone poles, in the USA; and *Sporothrix* against *Oidium* of hothouse cucumber or against fungi that cause wood to turn blue). We can also cite the sophisticated method of biocontrol of canker in chestnut trees by application of hypovirulent colonies of *Endothiella*.

Fusarium oxysporum ssp. *cannabis* is used as a **bioherbicide** for suppressing marijuana plants. In a more general way, *Colletotrichum gloeosporioides* and *C. truncatum* are also used as bioherbicides.

Harmful effects of Deuteromycetes

Medicine and veterinary medicine

In the field of medicine and veterinary medicine, Deuteromycetes are significant because of immunodepression in cases of transplants. The defence mechanisms of the organism are considerably diminished, and various microorganisms that are normally harmless become pathogenic.

— **Profound mycoses:** pulmonary aspergillosis (*Aspergillus*), lymphatic sporotrichosis (*Sporothrix schenckii*), equine lymphatic histoplasmosis (*Histoplasma*).

— **'Superficial' mycoses:** onychomycosis (*Scytalidium, Graphium*), chromomycosis (*Phialophora), ringworm (dermatophyte), fusariosis in crayfish (Fusarium*).

— **Allergies:** pulmonary, epidermal. Among a group of 15 fungi presently tested in this field, 11 are Deuteromycetes. They are all the more noxious as their conidia are carried, in houses, by animals such as the acarid *Tyrophagus putrescentiae*. Occupational pulmonary diseases: cheese industry (*Penicillium roquefortii*), breweries (*Aspergillus clavatus*), mushroom farms (*Doratomyces*). Skin allergies of cane harvesters of Provence (*Arthrinium*). Respiratory allergy in public parks in certain years (*Cryptostroma corticale*).

— **Fungal toxins:** these are numerous and induce various sicknesses in humans and animals. The best known are aflatoxins (produced by various species of *Aspergillus* and *Penicillium*), some of which are powerfully carcinogenic.

Phytopathology

In the field of phytopathology, Deuteromycetes cause a very large number of plant diseases that manifest themselves in various symptoms:

— **Tissue necroses**, limited or extensive, on leaves, flowers, fruits, stems, and roots (apple scab from *Spilocaea pomi*; anthracnose of plane tree from *Discula nervisequa*; peach shot hole from *Stigmina carpophylla*; canker of chestnut tree from *Endothiella* sp.; seedling wilt from *Fusarium* spp.; black spot of rose from *Marssonina rosae*; blossom blight of cherry tree from *Monilia laxa*; green and blue mould of citrus fruits from *Penicillium digitatum* and *italicum*; black rot of roots of numerous plants from *Chalara elegans*; needle cast of pines from *Lophodermium seditiosum*; silver scurf of potato from *Helminthosporium solani*).

— **Vascular wilting** (dutch elm disease from *Graphium ulmi*; verticilliosis of numerous herbaceous and woody plants from *Verticillium albo-atrum* and *V. dahliae*).

Alterations

Practically all foods and organic matter can be altered to the point of toxicity by Deuteromycetes. But their ravages extend to non-organic matter as well: for example, glass is damaged by *Penicillium citrinum*, metals such as aluminium and steel by *Aspergillus* and *Trichoderma*, and paints by *Phoma violacea*.

HOW TO USE THIS BOOK

You know little or nothing about the Deuteromycetes, or you wish to update your knowledge about this group.

- Definitely read chapter I, essentially from pp. 1 to 24. The basic concepts and terminology are explained in these pages (and the terms can be found again in the glossary).

You know the group, but you have under the microscope a member of the group in which you do not know the mode of conidiogenesis.

- The essential point is to determine how the fungus produces its conidia (conidiogenesis). Chapter I will help you, pp. 24–35, Figs. I.16 to I.30 and Table I.H.

You know the mode of conidiogenesis, or chapter I serves to clarify it.

- Refer to the corresponding chapter (which will enable you to confirm the character), then use the identification keys and the tables to arrive at the identification of the organism.

You do not know the meaning of a term.

- Consult the glossary, which often indicates as well where one can find an illustration of the term in the body of the work.

You wish to know what a given genus looks like, and/or you seek some information about it.

- The alphabetic index of taxa will lead you to the illustration of the genus. Brief information about it can be found here (number of species, geographic location, mode of life and substrate; then, if need be, the name of the teleomorph, recent bibliography in short form, existence of molecular study).

After finding information concerning a genus, you look up a bibliographic reference in short form that is difficult to understand.

- Look up the list of abbreviations used in the additional references.

Note

The **references to tables** within the keys to identification do not repeat the Roman numeral of the chapter, e.g., **C2, C4** (except in chapters containing a single table). It is only when we refer to a table outside the chapter that we specify the chapter number (e.g., VII A 12).

In the **body of the work**, we use the terms 'conidia' and 'spores' interchangeably when we speak of the mitotic conidia, which are asexual spores, of Deuteromycetes.

Similarly, some **abbreviations** of current usage are used here: SEM (scanning electron microscope), TEM (transmission electron microscope), LM (light microscope).

Regarding the **legends** accompanying the tables illustrating the genera, we have combined a certain number of data available at the time of publication of this work:

— in particular, the additional references mention selective studies that are not found in the general bibliography;

— in the column 'teleomorphs' the indicated sexual stage(s) are noted for at least one of the species of the anamorph genus (which does not imply that all the species of the latter are linked to this teleomorph genus or genera).

— most of the teleomorphs of the Deuteromycetes belong to the Ascomycetes; a B following the name of the genus indicates that it belongs to the Basidiomycetes.

— in the last column, 'Bio. mol.', the presence of an X indicates the existence of a study or studies of at least one species of the genus considered with the help of molecular biological techniques.

Etienne Kiffer, the illustrator, has designed the **illustrations** to give a synthesized view, which is rather general and diagrammatic, of the genera.

I

INTRODUCTION

The Deuteromycetes or Mitosporic Fungi (Fungi Imperfecti) make up the higher fungi, with Ascomycetes and Basidiomycetes.

The higher fungi are **eukaryotic** organisms (organisms having a true intracellular nucleus, limited by a double membrane) that have chitinous cell walls. Their vegetative system is a thallus that is well developed, filamentous (= **mycelium**), and haploid or dikaryotic (cf. infra). They produce only non-motile spores (not zoospores that move by means of flagella). They are **heterotrophic**, that is, dependent for their energy needs on pre-formed organic matter that is dead (**saprophytism**) or living (**parasitism, symbiosis**).

The Deuteromycetes constitute an artificial group. They comprise:

— higher fungi that can live and multiply seemingly without a sexual phase, and

— most of the asexual or 'imperfect' forms (**anamorphs**) of Ascomycetes and Basidiomycetes. (Arbitrarily, the anamorphs of Basidiomycetes that cause rust (Uredinales) and smut (Ustilaginales), parasites of vascular plants, are not included in Deuteromycetes.)

Among the Ascomycetes and Basidiomycetes, **sexual reproduction** takes place during a stage of the biological cycle called the 'perfect stage' or **teleomorph**.

During their sexual phase, the Ascomycetes produce reproductive cells, the **ascospores**, in parent cells or **asci**.

The main stages in the life cycle of an Ascomycete are shown in Fig. I.1a and occur as follows:

The **haploid** ascospores (having a nucleus with n chromosomes), released when the ascus matures, germinate when they are in a substrate with conditions favourable for the formation of a vegetative mycelium with haploid, uninucleate hyphae.

Then, two mycelia of different polarity undergo a fusion of their cytoplasm (**plasmogamy**) while the parental haploid nuclei remain, separately, in a common cytoplasm. This plasmogamy can occur in various ways: the fusion of an **ascogonium** (female sex cell) and a **spermatium** (male gamete) is shown in Fig. I.1a.

It is at the end of the dikaryotic phase (*di* = double, *karyon* = nucleus), characterized by a binucleate mycelium (reduced to an ascogenous filament), that the ascogenous cell is formed. At this point the haploid nuclei fuse (**karyogamy**) to form a diploid nucleus (a nucleus with 2n chromosomes).

In the ascogenous cell thus transformed into an ascus, meiosis occurs, which results in four haploid nuclei, around which the ascospores are differentiated (numerous variations exist around this type of base, particularly by abortion or division of the meiotic nuclei. The number of ascospores thus varies from 1 to *n*, and most commonly it is 8).

A comparable sexual reproduction occurs among the Basidiomycetes: in this case, the parent cell is the **basidium**, which produces exogenous **basidiospores** (Fig. I.1b). When they germinate, these basidiospores produce a haploid or **primary** mycelium. The fusion of two primary mycelia of different polarity produces a dikaryotic or **secondary** mycelium, which may remain in the vegetative stage for a long time, unlike what happens with the Ascomycetes (cf. supra).

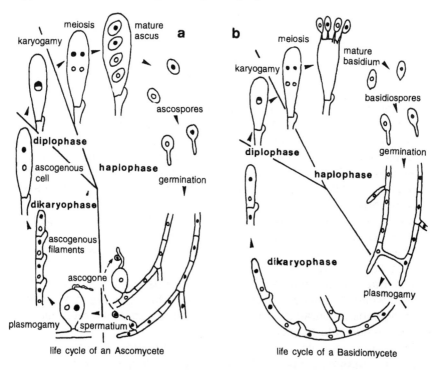

Fig. I.1. Life cycles of the higher fungi.

The Deuteromycetes, in contrast, are characterized by **asexual multiplication** by **mitosis**, considered secondary to sexual reproduction, from which the name of this group (*deuteros* = secondary, *mukes* = fungus) is derived. Most often, this multiplication occurs by production of mitotic spores or **conidia**, from a specialized hypha called **conidiophore** (Fig. I.4). The conidiophore is formed from a mycelium produced by dispersal units (ascospore, basidiospore or conidium, cf. Table I.Aa). The conidiophores are single in **hyphal forms**, or grouped on or in various structures: the **fruit bodies**, in **conidiomal forms**. The production of conidia, or **conidiogenesis**, occurs by various means and on various structures (cf. Table I.H at the end of the chapter). These enable us to classify and identify them, and constitute the subject of this work. Certain Deuteromycetes, however, do not produce conidia, but only mycelial forms or forms of resistance (Table I.Ad, Fig. I.29). These are the *Mycelia Sterilia* or Agonomycetes discussed in chapter XIII.

In some cases, in the life cycle of a fungus, a phase of sexual reproduction (teleomorph) can coexist with a phase of asexual multiplication

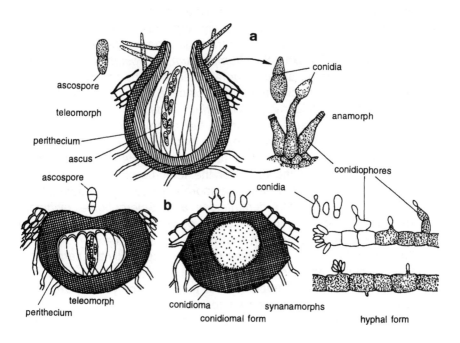

Fig. I.2. Examples of teleomorph-anamorph alternation.
a: life cycle of an Ascomycete (*Venturia inaequalis*, teleomorph) with its anamorph (*Spilocaea pomi*); **b:** an Ascomycete (*Sydowia polyspora*, teleomorph) with two anamorphs (synanamorphs) in its life cycle, a conidiomal form (*Sclerophoma pythiophila*) and a hyphal form (*Hormonema dematioides*).

Teleomorph, anamorph, holomorph

There are several possibilities:
— The fungus cycle is complete and described, that is, the characteristics and relationships of teleomorph and anamorph are established.
— The teleomorph and anamorph(s) are described separately, but their relationships are not established.
— Anamorph does not exist or is not described.
— Teleomorph does not exist or is not described.
— The same fungus can produce two (rarely several) different anamorphs or **synanamorphs** (Hughes, 1979), e.g., Fig. I.2b.

The terms 'teleomorph', 'anamorph', and 'holomorph', used hereafter, are given by Hennebert and Weresub (1977). These terms were questioned by Sutton (1993), who proposed to replace the first two with, respectively, meiosporic fungus and mitosporic fungus. This proposal was criticized by Korf and Hennebert (1993).

(anamorph): see, for example, Fig. I.2a. In such cases, the term **holomorph** is used to designate the fungus considered in all its phases, including teleomorph and anamorph (see box).

Etymology: *ana* = from the bottom up (development phase), *teleos* = end (end phase), *holos* = entire (complete), *morph* = form.

As regards nomenclature, the two types, teleomorph and anamorph, have generally been studied separately and thus have been given different **binomials** (*binomial* = a term consisting of generic name and species adjective). For example, *Stemphylium botryosum* is the anamorph of *Pleospora herbarum*, an ascosporate teleomorph. The two forms can also have the same species name, for example, the anamorph *Pollaccia mandshurica* and its teleomorph *Venturia mandshurica*, which are otherwise described together. More rarely, the anamorph does not have a specific name; we say, for example, 'the anamorph *Dicyma* of *Ascotricha distans*'. When the life cycle of a fungus is complete and known, the holomorph should have the name of the teleomorph, e.g., Fig. I.2b, in which the fungus comprising the teleomorph *Sydowia polyspora* and the synanamorphs *Hormonema dematioides* and *Sclerophoma pythiophila* has the name of the teleomorph *Sydowia polyspora*. It is only when referring specifically to an anamorph that one uses its name.

The sexual forms are sometimes massive, especially the carpophores of Basidiomycetes, and also of certain Ascomycetes.

The asexual forms, characteristic of the Deuteromycetes, are on the contrary generally small or even microscopic.

The teleomorphs of Deuteromycetes are most often Ascomycetes, and more rarely Basidiomycetes.

We mention later, in the introduction to each chapter, and along with each generic table, the known links between anamorphs and teleomorphs, as drawn from the compiled references, i.e. Kendrick (1979), Sutton and Hennebert (1994). With regard to the Ascomycetes cited in these lists, we have adopted the nomenclature proposed by Eriksson and Hawksworth (1993).

Even if the teleomorph of a Deuteromycete is unknown, it can sometimes be established as belonging to Ascomycetes or Basidiomycetes, particularly by the characteristics of the septa of the filaments or **hyphae**. They are simple in the Ascomycetes, and have a central pore that can close up if necessary by means of a special organelle called **Woronin body** (Fig. I.3a). In the Basidiomycetes, the pore, with a bulging edge, is named the **dolipore**: on each side of it is a hemispherical **parenthesome** that issues from the endoplasmic reticulum (Fig. I.3b). The study of this characteristic requires the use of an electron microscope. But one can also establish this relation by techniques of molecular biology.

According to Muller (1981), the anamorphs of Ascomycetes are always haploids, while those of Basidiomycetes can be haploid (primary mycelium) or dikaryotic (secondary mycelium). In the latter case, one can often observe clamp connexions characteristic of secondary mycelium (Table I.Ag).

Table I.A. Stages of the life cycle

a: dispersal unit: basidiospore, ascospore (produced by the teleomorph) or conidium (produced by the anamorph); **b:** germination of the dispersal units; **c:** mycelium arising from the germination; **d:** mycelium not producing conidia (chapter XIII); **e:** production of conidia by conidiophores (chapters II-XII); **e1:** single conidiophores carried by the mycelium; **e2:** conidiophores grouped in a fruiting body arising from the mycelium; **f:** simple mycelium of Ascomycetes and certain Basidiomycetes; **g:** dikaryotic mycelium with clamp connexions of certain Basidiomycetes; **h:** basidium with exogenous basidiospores; **i:** ascus with endogenous ascospores; **j:** unicellular yeasts (multiplication without production of mycelium).
N.B. The groups in the box (**h, i, j**) are not dealt with in this work.

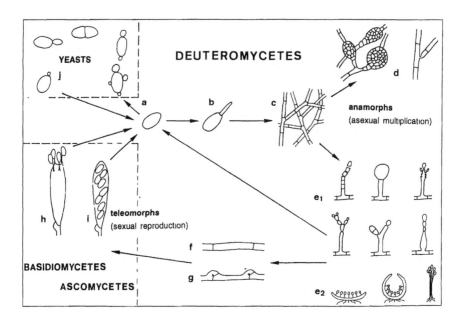

Other fungus groups, apart from Ascomycetes and Basidiomycetes, have anamorphs, but they are not grouped among Deuteromycetes. They are mainly Oomycetes and Zygomycetes. The unicellular yeasts, which multiply by budding or fission (Table I.Aj), are not discussed here.

Table I.A attempts to summarize and group the ideas that we will discuss and especially to place the Deuteromycetes in the life cycles of the higher fungi.

The different classifications

The taxonomy of Deuteromycetes is artificial: it does not constitute a system of classification such as a genealogical tree representing the natural relationships of these fungi, but rather a nomenclature and means of identification. From this perspective, we adopt here a taxonomy that uses inputs from different phases in the evolution of knowledge.

The first coherent system of classification of Fungi Imperfecti was proposed by Saccardo (1880, 1884). It was a morphological system, taking into account:

— the mode of grouping of conidial apparatus (Table I.B) in Hyphomycetes, Coelomycetes, etc.;

— the colour, form, and septation of the conidia produced (Table I.C).

Such a purely morphological system leads us to bring together Fungi, which a careful observation reveals as producing their conidia by different mechanisms: such observations were done on a small number of samples till the middle of the twentieth century.

In 1953, Hughes proposed a coherent, generalized classification of the Hyphomycetes. He divided them into eight sections (the suprageneric sections of Saccardo, families, etc. were abandoned and the sections designated simply by an order number). The 'characters of conidiophore and conidium development' are the basis of this classification (Table I.D).

This publication gave rise to a great number of works and some of them are cited in the bibliography at the end of this chapter. The sections of Hughes were given names and Barron (1968), whose system we generally follow, established the equivalents given in Table I.E. We have also indicated in this table the illustrations given in chapter I, and the chapter that discusses this group, followed by the name we have adopted.

Conidial ontogeny, the relationships between the conidium and the conidiogenous cell, evolution of the conidium and conidiophore, and the mode of secession of spores—the bases of modern classification—must be examined very carefully, especially at the level of the walls (see box). The light microscope is insufficient for the study of these phenomena, and the electron microscope is used for a certain number of observations. We have proposed interpretations of published photomicrographs, partly in this introduction, and then at the beginning of each chapter discussing

Unitary parameters of conidiogenesis

Hennebert and Sutton (1994) proposed 20 parameters, the combination of which should enable us to characterize perfectly the genera and species of anamorphs. The authors try to propose the largest number of parameters possible in each case, and if this system is adopted, each genus and species described or studied must be characterized according to these criteria.

Table I.B. Fungi Imperfecti.
Classification based on the mode of grouping and the
pigmentation of spores (mainly from Saccardo and Grove).

HYPHOMYCETES	AGONOMYCETALES	sterile mycelium		AGONOMYCETACEAE
	HYPHOMYCETALES	conidiophores dispersed in the substrate		DEMATIACEAE (dark colour)
				MUCEDINACEAE (hyaline or brightly coloured)
	STILBELLALES	conidiophores grouped in coremia		STILBACEAE
	TUBERCULARIALES	conidiophores grouped in a globular stroma (sporodochium)		TUBERCULARIACEAE
COELOMYCETES	MELANCONIALES	conidiophores grouped in a thin, submerged stroma (acervulus)		MELANCONIACEAE
	SPHAEROPSIDALES	conidiophores grouped in a complete pycnidium		SPHAERIOIDACEAE (dark colour)
				NECTRIOIDACEAE (hyaline or brightly coloured)
		conidiophores grouped in an incomplete pycnidium		LEPTOSTROMATACEAE
		conidiophores grouped in a pseudo-pycnidium in the shape of a dish		EXCIPULACEAE

Table I.C. Types of spores (according to Saccardo)

Morphology	Definition	Type	Colour	
			hyaline or clear	black or dark
	unicellular, globular or slightly elongated	Amerospores	Hyalosporae	Phaeosporae
	bicellular	Didymospores	Hyalodidymae	Phaeodidymae
	multicellular, transverse septa	Phragmospores	Hyalophragmiae	Phaeophragmiae
	muriform; transverse and longitudinal septa	Dictyospores	Hyalodictyae	Phaeodictyae
	unicellular or septate very elongated	Scolecospores		
	unicellular or septate, spiral or helicoid	Helicospores		
	unicellular or septate, star-shaped or more or less irregular	Staurospores		

These definitions apply in principle to all types of spores, including conidia.

a group. For the category of Blastosporae *sensu lato* (chapters IV to IX), the interpretations are combined in chapter III.

The taxonomy adopted: principles

In this work, which we hope to make practical and pedagogical, we adopt a hybrid taxonomy, taking what appears to us to be useful from the different systems discussed above.

    — From the classifications of Saccardo (1880, 1884) and Grove (1919), we retain the study of the mode of grouping of conidiophores (Table I.B). But we give it a secondary importance, and we choose sections that are different from those of the authors in defining the two major parameters for grouping:

• separate conidiophores: hyphal form (Fig. I.4)

• conidiophores grouped in conidiomata: conidiomal forms (Table I.F).

The term 'conidioma' (Kendrick and Nag Raj, 1979) corresponds to basidioma and ascoma, designating the fruit bodies of Basidiomycetes and Ascomycetes. We include here the coremia of Stilbellaceae, the sporodochia of Tuberculariaceae, the acervuli of

Table I.D. Classification of Hughes (1953). Terminology proposed by the author. **IA:** blastospores; **IB:** solitary blastospores (above), botryoblastospores solitary and in chains (below); **II:** terminal spores; **III:** solitary chlamydospores and successive ones on annellophores; **IV:** phialospores; **V:** meristem arthrospores; **VI:** porospores; **VII:** arthrospores; **VIII:** basauxic conidiophores.

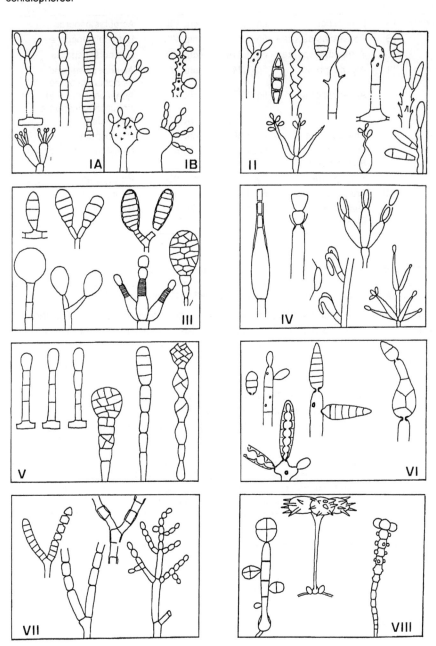

Table I.E. Correlations between the taxonomies of
Hughes, Barron and the present work

Hughes, 1953		Barron, 1968	Illustrations	Chapters	Names adopted
Section	IA	Blastosporae	fig. I.21a	ch. VII	Acroblastosporae
	IB	Botryoblastosporae	fig. I.22	ch. VIII	Botryoblastoporae
	II	Sympodulosporae	fig. I.21b	ch. VI	Sympodulosporae
	III	Aleuriosporae	fig. I.24	ch. IV	Aleuriosporae and Monoblastosporae
		Annellosporae	fig. I.27a	ch. V	Annellophorae and Annelloblastosporae
			fig. I.27b	ch. XI	Annellidae
	IV	Phialosporae	fig. I.26	ch. X	Phialosporae
	V	Meristem-Arthrosporae	fig. I.18	ch. II	Meristem Arthrosporae
	VI	Porosporae	fig. I.23	ch. IX	Porosporae
	VII	Arthrosporae	fig. I.17	ch. II	Arthrosporae
	VIII	Meristem-Blastosporae	fig. I.28	ch. XII	Basauxic Deuteromycetes

Melanconiaceae, the pycnidia and pseudopycnidia of Sphaerioidaceae and Nectrioidaceae, the thyriopycnidia of Leptostromataceae and the cupuliform fructifications of Excipulaceae, of Saccardo's classification (Tables I.B and I.F).

Of these, the criteria of form, septation and spore coloration (defined in Table I.C) are retained but at a different level, most often generic, or even specific (a genus can include, for example, amerosporate, didymosporate, phragmosporate and dictyosporate species, or species with hyaline (colourless) conidia, or conidia that are more or less coloured).

— With Hughes and his successors, we take spore ontogeny as the basis of classification. In consequence, the groups of Deuteromycetes are defined by this criterion (Tables I.E and I.H) and examined as such, hyphal and conidiomal forms together, in the chapters that follow.

The system of Saccardo is outdated primarily because it does not take account of this last criterion, and also because it is hardly applicable in culture, where the differences between acervulus, sporodochium, and cupuliform fructification become blurred or even disappear (the conidioma becomes atypical and often the dispersed conidiophores come together, in the manner of the Hyphales). This is why a species of Coelomycetes should never be described simply from a culture (Nag Raj, 1981).

In consequence, certain genera are found many times in our key, according to the forms in which they occur.

The criterion of spore ontogeny has primarily been described in the 'Hyphomycetes'. It is sometimes inadvisable to apply it to 'Coelomycetes' because of difficulties in observation.

As a general rule, the reader will seek to see how the fungus produces its conidia: that will lead him to the chapter where the group thus characterized is described, and which contains a key and synoptic tables for the identification of genera.

The taxonomy adopted: definitions

The fungal wall (Fig. I.3).

The anamorphs of Ascomycetes do not present a particular cytology, and it is necessary to study the peculiarities of spore ontogeny, essentially at the level of the walls of the conidiogenous cells and the conidia. It is thus important to define clearly the constituents of this wall, and we follow the terminology of Reisinger (1972) and Reisinger *et al.* (1977a) in distinguishing layer A and layer B, from the exterior towards the interior (Fig. I.3a).

Layer A

Generally, layer A forms an amorphous, granulous pellicle in aerial organs. The state of its surface determines the ornamentation of the fungal

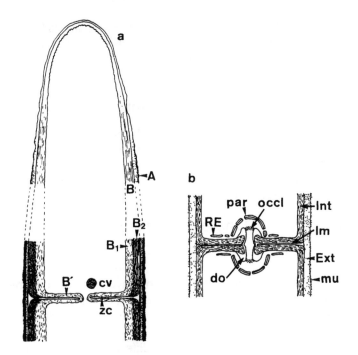

Fig. I.3. Walls and septa of higher fungi.
a: wall and septum of hyphae of Ascomycetes; **A:** layer A; **B:** layer B; **B1:** non-melanized B; **B2:** melanized B; **B′:** B of the septum; **zc:** zone of cleavage; **cv:** Woronin body, organelle that closes the pore in the septum; **b:** wall and septum in the Basidiomycetes, Hymenomycetes. **Ext.:** outer layer; **Int.:** inner layer; **lm:** middle lamella; **do:** dolipore, pore with a bulging edge; **occl:** occlusion of the pore; **par:** parenthesome, hemispheric organelle arising from the endoplasmic reticulum; **RE:** endoplasmic reticulum; **mu:** external mucilage.

elements (Fig. I.10). The pellicle is absent and generally replaced by a mucilaginous zone when the fungal organ grows through the substrate or in a liquid.

Chiefly polysaccharidic in nature and possibly charged with pigments (melanins), layer A appears to be simply an exudation of the underlying layer B.

Layer B

Layer B is the fundamental layer of the wall of all fungal organs. It is developed by the hyphal apex or the conidial primordium. It is partly polysaccharide, partly glycoprotein and chitin, fibrous. In pigmented fungi, one can distinguish two zones: B2, the external layer, melanized, arising from B1, the internal, transparent layer. The importance of the two zones varies. Certain organs can finally possess a wall that is entirely melanized.

(Layer C exists only in certain conidia; it can be seen later, in the structure of the conidial walls, Figs. I.8 and I.9.)

The Basidiomycetes seem to have a comparable wall organization, but the transverse septa show pores that are more or less complex (Khan and Kimbrough, 1982). Figure I.3b shows the dolipore with parenthesomes of a Basidiomycete, Hymenomycete.

Conidial apparatus

Except in the case of a fungus that remains totally sterile, in Deuteromycetes there is production of **conidia**, spores that are asexual, mitotic, exogenous, and non-motile. They are produced by the **conidiogenous cells**, possibly carried by a **conidiophore**. All these constitute the conidial apparatus (Fig. I.4).

The conidiophores

Type

The conidia can be formed directly on the mycelium = **micronematous** conidiophore (Fig. I.4a, b). One can also consider that there is no conidiophore at all, especially in case 4a, where a pre-existing hypha is transformed into conidia.

When the conidiophores are poorly differentiated, but tend to keep conidia away from the vegetative mycelium, they are called **semi-macronematous** (Fig. I.4c, d).

Finally, when the conidiophore is very distinct from vegetative hyphae, erect, and differentiated in several ways, it is called **macronematous**. It is typically made up of the stipe, the sterile branches, and the conidiogenous cells (Fig. I.4e).

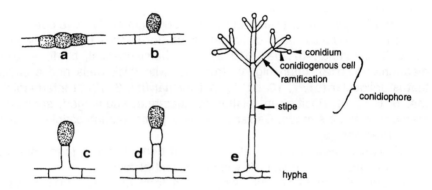

Fig. I.4. Different types of conidiophores.
a, b: micronematous conidiophores; **c, d:** semimacronematous; **e:** macronematous.

Conidiogenous cell

The conidiogenous cell is more or less merged or integrated with the conidiophore (Fig. I.4a, b, c) or on the contrary individualized on a structure that is more or less complex (Fig. I.4d, e).

Moreover, its function evidently determines the mode of production of conidia, otherwise called conidiogenesis. This fundamental point is examined elsewhere (conidial secession, different types of conidia and groups of Deuteromycetes).

Mode of grouping

The isolated conidiophores (hyphal forms) are called **mononematous**. The conidiophores grouped in **conidiomata** (Michaelides in Kendrick and Nag Raj, 1979) (conidiomal forms) are grouped in various ways:

— *inside a conceptacle* (reproductive organ, more or less enclosed, with walls containing conidiophores and conidia).

Pycnidium (Table I.Fa): Fructification develops from one or several hyphae, globular to lageniform, isolated or gregarious, variously pigmented, with a cavity normally undivided, and neither convoluted nor multilocular, with generally thin walls. It is closed (a1) or provided with a circular ostiole at the tip, which may be papillate (a2) or rostrate (a3). The pycnidium may be submerged (sometimes provided with a **clypeus**, shielded with fungal tissue developed from the upper part of this, a4), semi-submerged, or superficial, in the last case with (a5) or without a **subiculum** (fungal tissue that is more or less loose at the base of the pycnidia). It is generally glabrous and sometimes has brown, multicellular bristles or short, hyaline, aseptate bristles. It is not rare to find a culture with multiostiolated pycnidia.

Pseudopycnidium (Table I.Fb): fruit body generally developed from a stroma, a compact mass of fungal tissue, for various forms (pulvinate, b1 to 5; columnar, b7; flat, b10 and 11; hemispherical, b1 to 4, b11; meandriform, b12), often highly stromatic, with thick walls and a cavity that is unilocular (b1, 7, 10, 11) to multilocular (b2, 3, 12) or intermediate (b4, 5, 6, 8, 9). Position in relation to substrate: submerged, semi-submerged, or at the surface. Dehiscence takes place through ostiole, fissure, or an irregular gap.

Thyriopycnidium (Table I.Fc): flat fruit body, but in a form shielded by a radiated structure with or without a central foot.

Cupule (Table I.Fd): fruit body in the form of a disc (d1) or dish, most often at the surface, opening fully at maturity, often ornamented at the periphery with sterile hyphae or bristles that are pigmented or hyaline and variably compartmentalized.

— *at the surface of a receptacle* (fructiferous organ having conidiophores and conidia on its surface).

Acervulus (Table I.Fe): fruit body consisting of a basal stroma that is poorly developed and pseudoparenchymatous (with non-discernible hyphae), covered with the host tissue. No walls. Conidiogenous cells limited at the floor of the cavity. Dehiscence by rupture of the host tissue. Dark brown bristles are sometimes present, either among the conidiogenous cells or at the edge of the acervulus. The latter can be subcuticular (e1), intraepidermal (e2), subepidermal (e3), or subperidermal (e4).

Sporodochium (Table I.Ff): fruit body in the form of a large or pustuliform stroma, superficial or greatly erumpent (f1). Related to these are the pulvinate forms with smaller stroma (f2) or tufted conidiophores (f3).

Coremium (Table I.Fg): conidiophores in a compact mass rising in a small column. The fertile zone may be limited to the apical part (capitulum, g1) or may occupy the entire coremium (g2). There are also compound coremia in which the principal axis has lateral, fertile branches (g3).

We wish to stress (see box p. 17) that a fruit body must necessarily be studied from a longitudinal section, not only for characterizing the type, but also for judging its position in relation to the substrate, the distribution of conidiophores and conidiogenous cells, the type of conidiogenesis, and finally the texture of the conidioma itself.

Texture of conidiomata

The **plectenchymatous** mycelial mass (as all fungal tissue is described) constituting the conidiomata has various textures, which may be used for

Table I.F. Fruit bodies

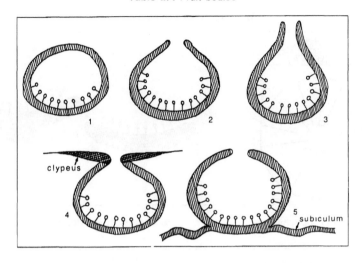

a. pycnidium: 1, closed; 2, with papillate ostiole; 3, with rostrate ostiole; 4, attached to a clypeus; 5, on a subiculum

b. pseudopycnidium: 1-12, variety of forms and septations (see text)

Table I.F. Fruit bodies (contd.)

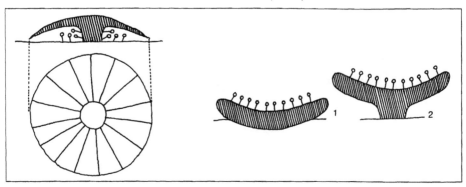

c. thyriopycnidium   **d. cupule:** 1, sessile, 2, stipitate

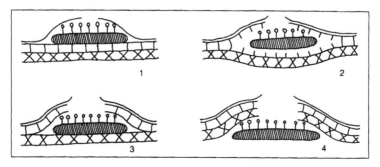

e. acervulus: 1, subcuticular; 2, intra-epidermal; 3, subepidermal; 4, subperidermal

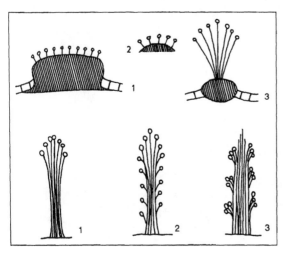

f. sporodochium:
1, typical; 2, reduced, pulvinate;
3, reduced, with caespitose
conidiophores

g. coremium:
1; fertile at the tip; 2, entirely
fertile; 3, with fertile branches

Variability of conidiomata

Intermediaries may be seen between the types defined above: thus, the cupulate coremia defined by Seifert (1985), an example of which Roquebert and Bury (1988) studied in detail in *Leptoxyphium* sp., show analogies with stipitate pseudopycnidia (Table I.F, b7).

This diversity was analysed by Okada and Tubaki (1986), who showed the existence of a continuum between the different types of coremia and even the passage towards other aggregate forms, such as the sporodochia and the pycnidia. If the sporodochium is vertically developed it can be thought to be a coremium, but its *textura angularis* distinguishes it. According to Seifert (1985), this texture does not exist in true coremia.

But above all the pulvinate forms (Table I.F, f2), in the same manner as bundles of more or less defined hyphae, enable us to link together hyphal and conidiomal forms.

---

the purpose of generic determination. These different types of textures (defined in Table I.G) were recognized for the first time in Ascomycetes (Starback, 1895; Korf, 1952).

The conidia

We will look successively at various characteristics that enable us to differentiate the types of conidia.

Thalloconidia and true conidia

The first distinction is made between **true conidia**, produced *de novo* from a specialized conidiogenous cell (Fig. I.5a), and **thalloconidia**, formed from a pre-existing part of thallus (Fig. 1.5b). At the Kananaskis symposium (Kendrick, 1971a), it was specified that the thalloconidia or thallic conidia are limited by a septum *before* a possible increase in volume, while the true conidia or blastic conidia (arising from a process of budding) are limited by a septum *after* bulging.

Origin of the conidial walls

Since the first Kananaskis symposium (Kendrick, 1971a), great importance was given to the question of knowing whether the wall of the conidium is in continuity with that of its conidiogenous cell (Fig. I.6a)

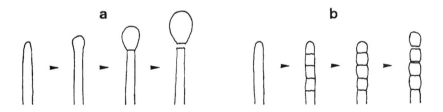

Fig. I.5. True conidia (**a**) and thalloconidia (**b**).

Table I.G. Different types of conidioma texture

Tissue made up of short cells: the hyphae are not discernible (= pseudoparenchymatous)

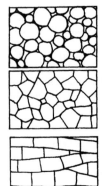

• Rounded to polyhedral cells

a rounded cells with intercellular spaces, *textura globulosa*

b polyhedral cells without intercellular spaces, *textura angularis*

• More or less rectangular cells

c *textura prismatica*

Tissue composed of elongated cells: the hyphae are easily discernible (= prosenchymatous)

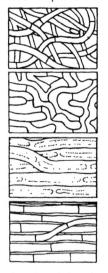

• Intermingled, non-parallel hyphae

d hyphae with walls that are not fused together, often with interhyphal spaces, *textura intricata*

e hyphae with walls that are fused together, without interhyphal spaces, *textura epidermoidea*

• Hyphae that are more or less parallel

f hyphae with thick walls and narrow lumina, *textura oblita*

g hyphae with thin walls and wide lumina, *textura porrecta*

(**holoblastic** conidia) or formed *de novo* inside the latter (Fig. I.6b) (**enteroblastic** conidia). It is often difficult to observe this character under the light microscope and interpret it (Hammill, 1974).

Septation and form of conidia

Saccardo (Table I.C) proposed to classify the spores according to their form and septation.

In the septate spores (phragmospores), two cases can be distinguished according to the order of appearance of the septa (Fig. I.7).

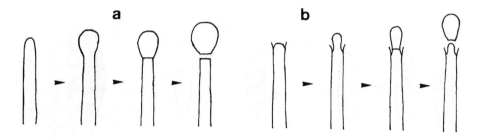

Fig. I.6. Holoblastic conidia (**a**) and enteroblastic conidia (**b**).

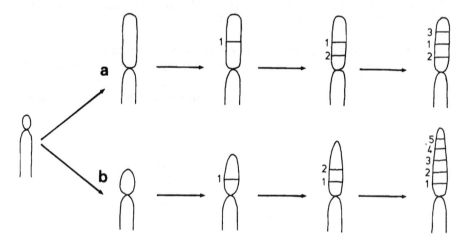

Fig. I.7. Types of maturation of phragmoconidia: (**a**) holospore; (**b**) acrospore.

Structure of the conidial wall

There are two types of conidia among the phragmoconidia, according to the structure of the walls and septa (Luttrell, 1963). The **euseptate** conidia have walls and septa that resemble hyphal cells (see above, the fungal wall, Fig. I.3a). **Distoseptate** conidia have false septa that are due to the juxtaposition of internal locules, which are released when the conidia break and emerge from the torn external wall (Fig. I.8b).

In fact, the distoseptate conidia possess a wall that is different from that of hyphae and other conidia (Fig. I.8c). On the interior of the fundamental layer B there is a supplementary layer C (sometimes subdivided into C1 and C2, numbered according to their order of appearance). It is fibrous and non-pigmented, and C1 appears mucilaginous (Reisinger et al., 1977a).

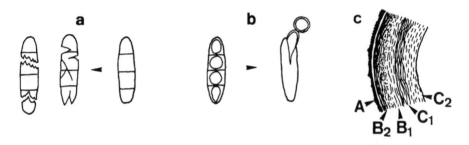

Fig. I.8. Euseptate conidia (**a**) and distoseptate conidia (**b**); ultrastructure of the wall of the latter (**c**): layers A, B (B1 + B2) and C (C1 + C2).

Certain conidia, of an intermediate type, have normal septa like the euseptate conidia, but a supplementary layer C is formed (Fig. I.9).

As layer C may also exist in non-septate conidia, Mangenot and Reisinger (1976) proposed the terms **haplothecate** and **diplothecate** to replace euseptate and distoseptate (Fig. I.9).

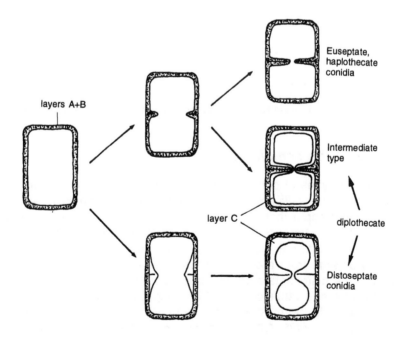

Fig. I.9. Haplothecate and diplothecate conidia.

Pigmentation and modifications of the conidial wall

Pigmentation
The fungal wall may or not contain melanin or other pigments, which can give conidia the following characteristics:
— hyaline: colourless or clear by transmitted light; white, or clear in mass by reflection;
— brightly coloured (not dark): green, yellow, pink, etc.;
— melanized: coloured brown, black, brown to dark green, etc.
The colourless and brightly coloured conidia characterize the old group of Mucedinaceae and the melanized conidia the group of Dematiaceae, in the system of Saccardo.

Ornamentation
The wall of conidia may be smooth (Fig. I.10a), echinulate (Fig. I.10b), or verrucose (warty; Fig. I.10c). In the last two cases, the surface is determined by layer A of the wall: the echinulations are peelings of layer A, full of pigments (Fig. I.10d), and the warts are larger and more or less empty (Fig. I.10e). More rarely, the wall may be striated, reticulate, etc.

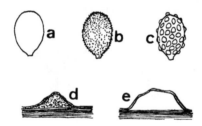

Fig. I.10. Types of ornamentation of the conidial wall: smooth (**a**), echinulate (**b**), verrucose (**c**), echinulation (**d**), wart (**e**).

More significant modifications can also be seen and are found in various forms:
— of seemingly acellular nature: cilia (Fig. I.11a–d), sheaths, or mucoid appendages of various forms (Fig. I.11e–i) located at different points on the body of the spore;
— of seemingly truly cellular origin: appendages simple (Fig. I.11j, l, m) or ramified (Fig. I.11k), also located at different points;
— finally, it is difficult to specify whether certain accessory cells of complex conidia are just cellular appendages such as those defined above, or functional elements (Fig. I.11n–p).

Germ pores and germ slits (Fig. I.12)
Punctiform or elongated zones are sometimes visible under the light microscope: these zones appear particularly clearly in coloured spores.

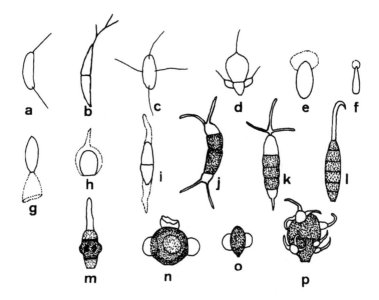

Fig. I.11. Secondary appendages and accessory cells of conidia. **a-d:** cilia; **e-i:** mucoid sheaths and appendages; **j-p:** cellular appendages.

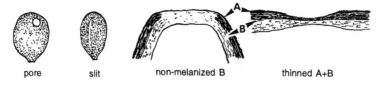

pore   slit   non-melanized B   thinned A+B

Fig. I.12. Germ pores and germ slits.

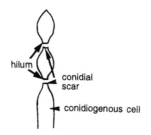

hilum
conidial scar
conidiogenous cell

Fig. I.13. Hilum and conidial scar.

Transmission electron microscopy shows that they correspond to a thinning of the wall or an absence of melanin in layer B. This results in lesser resistance that enables the emergence of the germ tube.

Hilum

The hilum is a scar, on the conidium, of the latter's insertion on the conidiogenous cell (Fig. I.13). The details of its structure depend on the mode of secession (cf. infra).

Conidial scar

The scar is the trace of the insertion of the conidium on the conidiogenous cell. A conidium having produced another conidium, therefore, itself has a conidial scar (Fig. I.13).

Conidial secession

The conidia must be released in order to be able to germinate and reproduce a new vegetative apparatus.

Two principal modes of release are known, schizolytic and rhexolytic.

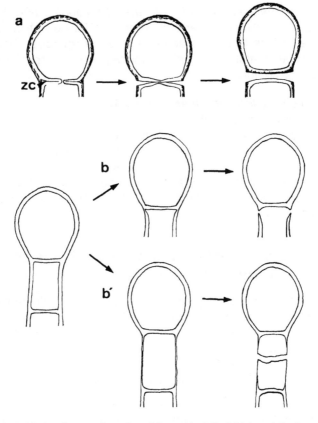

Fig. I.14. Mode of secession of conidia: schizolytic (**a**), rhexolytic (**b** and **b'**).

— schizolytic secession (Fig. I.14a):
The basal septum of the conidium splits at the middle layer or zone of cleavage (zc).

The septum may have a Woronin body that covers the septal pore, or it may close completely. These septa are called schizogenous.

— rhexolytic secession (Fig. I.14b-b'):
The wall of the conidiogenous cell becomes thin and fragile and breaks in order to release the conidium. There may be a well-defined zone of rupture (Fig. I.14b) or the lysis may occur all along the wall (Fig. I.14b').

> The mode of secession affects the appearance of the hilum and the conidial scar. In particular, the rhexolytic separation may leave a large or small parietal fringe at the base of the conidium.

> The release of the conidia is most often passive, effected by various agents (air currents, water drops, animals, etc.). More rarely it is active, with a mechanism for projection. The characteristics of the spore surface, dry or mucous, ornamented or not, affect its mode of dispersal.

Arrangement of conidia

Position of the conidium in relation to the conidiophore (Fig. I.15a)
The conidium is called **acrogenous** when it is located at the top of the conidiophore. If it is located laterally, it is **pleurogenous**. Finally, if the conidia are located laterally as well as apically, it is an **acropleurogenous** arrangement of conidia.

Relation of conidia to each other (Fig. I.15b)
The conidia may remain solitary.

If they are produced successively by the conidiogenous cell and form a chain, the youngest at the base, the chain is **basipetal**.

If the conidia are produced themselves from other conidia, wherein the youngest are at the tip, the chain is **acropetal** (= basifugal). The acropetal chain may be simple or ramified, as opposed to the basipetal chain, which is always simple.

The taxonomy adopted: the different types of conidia and the groups of Deuteromycetes

In this section we will quickly review the different subdivisions of those Deuteromycetes for which we have retained the chosen names. They are examined in greater detail in the chapters devoted to each group, but here we can compare them.

Thalloconidia (thallic conidia)

Thalloconidia are conidia formed from a pre-existing part of thallus, from hyphal fragments modified to ensure the preservation and dispersal of the fungus.

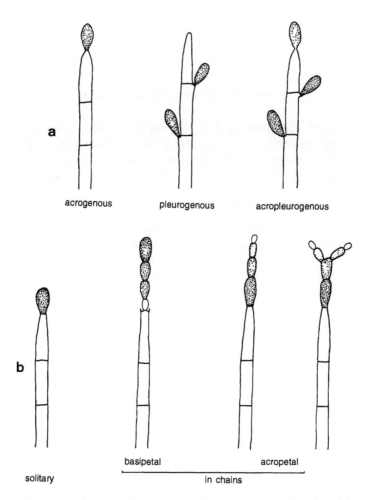

Fig. I.15. a: Positions of the conidia on the conidiophore; **b:** relation of conidia to each other; the acropetal chain can be simple or ramified.

Chlamydospores (Fig. I.16)

Found also in many fungus groups other than imperfect fungi, chlamydospores are generally not very characteristic and do not serve as a basis for taxonomy. They are typically portions of hyphae in which the cytoplasm is condensed: generally they are bulging and surrounded by a thick wall.

Arthroconidia

Resulting from the septation and disarticulation of a pre-existing hypha, arthroconidia are seen in all types from the most crude to the most evolved:

Fig. I.16. Formation of chlamydospores.

in *Geotrichum*, for example, it is actually the thallus, formed of prostrate hyphae, that develops into conidia (Fig. I.17a). In *Oidiodendron*, the branches of a well-differentiated conidiophore become segmented, bulge, and are surrounded by a more or less thick and ornamented wall (Fig. I.17b).

Meristem arthroconidia

The meristem arthroconidia result from the basipetal transformation of the conidiogenous hypha into conidia. Depending on the case, there may or may not be intercalary growth of this conidiogenous hypha (Fig. I.18). Here there is an apical bulging prior to septation, and this type of conidium is thus intermediate between thalloconidia and true conidia.

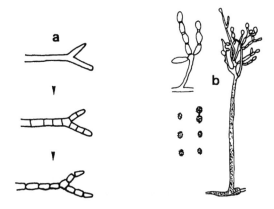

Fig. I.17. Examples of arthroconidia. **a:** formed from the thallus (*Geotrichum*); **b:** formed on a conidiophore (*Oidiodendron*).

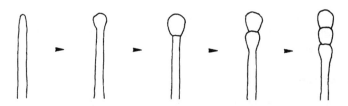

Fig. I.18. Formation of meristem arthroconidia.

Holothallic conidia (Fig. I.19)

Holothallic conidia are found essentially in Dermatophyta (parasites of the integuments and hair, nails ... of humans and animals) and result in the transformation of the tip of the parent cell, chiefly after it has attained its final length (Galgoczy, 1975). We name them Thalloaleuriospores and group them with the Aleuriospores, from which they are difficult to distinguish.

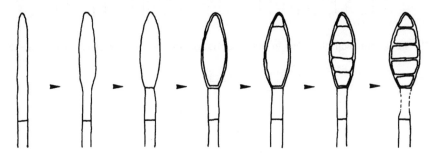

Fig. I.19. Formation of holothallic conidia.

True (blastic) conidia

Instead of arising like the thalloconidia from a pre-existing piece of thallus, true conidia are produced *de novo* by various mechanisms that we will examine, but that lead back to a process of budding, from which arises the name **blastic** conidia as opposed to **thallic** conidia (Kendrick, 1971a).

Blastosporae s.l. (sensu lato)

The production of conidia by budding is particularly clear in these related families (Fig. I.20).

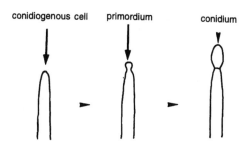

Fig. I.20. Formation of holoblastic conidia in Blastosporae *sensu lato*.

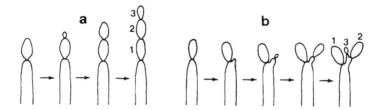

Fig. I.21. Formation of conidia in the Acroblastosporae (**a**) and the Sympodulosporae (**b**).

Fig. I.22. Simultaneous formation of botryoblastospores.

The wall of the conidium is continuous with that of its conidiogenous cell in the Blastosporae *s.l.*, which are thus holoblastic. We will now examine the different groups.

Briefly, the production of the first conidium is similar in Acroblastoporae and the Sympodulosporae, but after that the two processes differ (Fig. I.21).

Acroblastospores
In the Acroblastosporae, a new conidium (sometimes two or several) is produced by the preceding conidium and so on, giving rise to an acropetal (or basifugal) chain, the youngest conidium being at the tip (Fig. I.21a).

Sympodulospores
In the Sympodulosporae, there is a subapical regrowth of the parent cell and production of a new meristematic point, and the process is repeated, leading to the production of a bunch of single conidia on a conidiogenous cell at its more or less bulging or elongated tip.

Botryoblastospores
In the Botryoblastosporae, the conidia are produced simultaneously on a fertile, bulging or elongated head (Fig. I.22). They remain solitary or sometimes develop acropetal chains as in the Acroblastosporae.

Porospores (Tretoconidia)
This is an original creation of modern taxonomy: Hughes (1953) placed in section VI the Hyphomycetes, always melanized, forming their conidia

Fig. I.23. Annular thickenings melanized at the hilum and the conidial scar in the porospores (**a**: under light microscope; **b**: in TEM section).

'from pores on conidiophores'. This appears, under light microscope, as a melanized thickening surrounding a thin duct, at the site of the conidial scar; a pore is generally present also on the conidium, at the hilum (Fig. I.23a). It is from this appearance of the pore that the name of this spore type is derived.

> The porospores can remain solitary or bud in turn into other conidia to form chains: they result from a blastic process of budding, like the acroblastospores and the sympodulospores, from which they are not fundamentally different. For Reisinger (1972), the difference between blastospores and porospores lies in the maturation of the conidia and the evolution of the conidiophore (thickening and melanization of the wall around the pore, Fig. I.23b), not in the ontogeny, which is similar. In the case of certain genera (*Torula*, *Stemphylium*...), the authors hesitate to place them in either of these two groups.

Aleuriospores (hyphal) and monoblastospores (conidiomal)
These conidia are formed by the bulging of the tip of the conidiogenous cell, followed by the formation of a basal wall and often the thickening of the wall (Fig. I.24).

The aleuriospores are typically holoblastic, have a large base, and are solitary. Their definition is linked to the sporal ontogeny described above and the conidial separation may be rhexolytic or schizolytic.

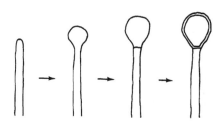

Fig. I.24. Formation of an aleuriospore.

Sutton (1980) distinguished in the 'Coleomycetes' a monoblastic form, of similar ontogeny, in which a unique conidium is produced on a parent cell that is generally not very differentiated. We use the term Monoblastosporae for this group.

Holoblastic annellospores of the Annellophorae (cf. discussion below on the annellospores)

Phialospores

Here also the process of budding of the conidia is clearly visible: the conidia are repeatedly produced at the tip of the conidiogenous cell or phialide (Fig. I.25).

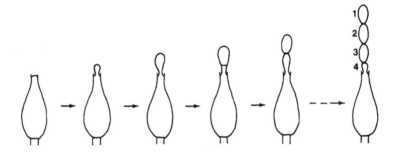

Fig. I.25. Formation of phialospores.

The meristematic point, situated at the tip of the phialide (= exophialide, Fig. I.26a) or more rarely at the interior of the collar (= endophialide, Fig. I.26b), is in principle fixed. If the conidia released from it remain attached to each other, they are progressively younger from the tip to the base of the chain, which is thus **basipetal** and non-ramified.

The phialospores are typically enteroblastic, since the conidial wall continues from the internal part of the wall of the phialide (Fig. I.26a).

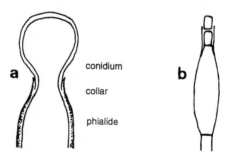

conidium

collar

phialide

Fig. I.26. Enteroblastic formation of the phialospore in an exophialide (**a**) and in an endophialide with a long collar (**b**).

Enteroblastic annellospores of Annellidae (cf. discussion below on the annellospores)

Annellospores

This subdivision was also made by Hughes (1953), who grouped it with the Aleuriosporae in section III. The Annellosporae are characterized by percurrent proliferations of the conidiogenous cell, which results in its elongation and in the presence of circular scars at its tip (Fig. I.27).

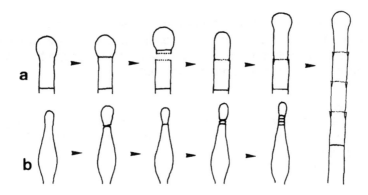

Fig. I.27. Difference between Annellophorae (a) and Annellidae (b).

This phenomenon was observed before Hughes but was not generalized as a taxonomic criterion (cf. Langeron and Vanbreuseghem, 1952).

It now clearly appears that this group is heterogeneous:

— it is characterized by its conidiogenous cells, not by the conidia;

— these conidia are of various origins and correspond to several types (Wang, 1990).

In some genera (e.g., *Acrogenospora*), they are aleuriospores that are holoblastic, rhexolytic (Fig. I.27a) or schizolytic; the regrowths of the conidiophore here are generally spaced out and few in number (Reisinger, 1972; Hammill, 1972b): Annellophorae.

In other genera (e.g., *Scopulariopsis*), they are enteroblastic phialospores, chiefly schizolytic, produced by the phialide, in which the collar elongates slightly because of the deposit of parietal material during the secession of each conidium (Fig. I.27b) (Kiffer et al., 1971; Hammill, 1971, 1972a, 1977): Annellidae.

An annellate aspect of the conidiogenous cell can be seen in a few Sympodulosporae; practically we have represented the genera concerned in the keys of the chapters on Annellidae, Annellophorae, and

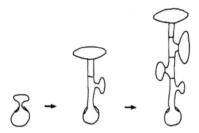

Fig. I.28. Growth of a basauxic conidiophore from the basal parent cell.

Sympodulosporae. We also include with Annellophorae the **Annelloblastosporae**, in which the grouping of conidiogenous cells in a conidioma makes it very difficult to observe under the light microscope the distinction between the annellophore and the annellide.

Other types
Certain Deuteromycetes present a particular type of conidiogenesis, among others the group of basauxic conidiophores (Fig. I.28). The conidia are produced by a conidiogenous filament put out in an enteroblastic manner by a basal, globular parent cell.

Mycelia sterilia or Agonomycetes

These are fungi that do not produce conidia: however, any non-fructified mycelium is not included in this category. The typical Agonomycetes have constant and recognizable structures that enable us to identify them. They are often seen in more or less massive aggregrates of cells such as sclerotia (with distinct cortex and medulla) and bulbils (without a differentiated cortex) (Fig. I.29).

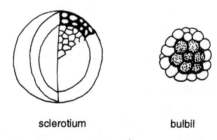

sclerotium   bulbil

Fig. I.29. Examples of propagules of Agonomycetes.

Conclusions

We have already mentioned that the group of Deuteromycetes constitutes probably a continuum (Minter et al., 1982, 1983a, b). The classifications in the group involve necessarily arbitrary divisions and make it difficult to place organisms in one or the other division. The problem for a mycologist studying the Deuteromycetes therefore consists in:

— recognizing the mode of conidiogenesis of the fungus being examined.

Let us recall that one must distinguish several stages: according to Reisinger (1972), initiation of primordium, conidial maturation, evolution of conidiophore.

Minter et al. (1983) themselves recognized up to seven 'critical stages' of the phenomenon: conidial ontogeny, conidial maturation, conidial delimitation, conidial secession, proliferation, regeneration, and collarette production.

— deciding the conidial group in which to place the organism before proceeding further with identification.

This is done most often by a **static** observation under **light** microscope, whereas the small size of the objects examined and the evolving character of conidiogenesis would necessitate **sequential** observations (Cole and Kendrick, 1968, 1969a; Kendrick and Cole, 1968, 1969; Kendrick et al., 1968) and the **electron** microscope.

A careful observation, deductions drawn from the examination of the fungus at various stages in the preparation, and the use of certain indices such as the nature of conidial chains (see the difference between the chains of the Blastosporae and the Phialosporae) generally enable the experienced observer to obtain sufficient information to effect identification.

It remains only to be said that one does not always know in which group to place certain genera: for example, those that produce solitary conidia on tapering conidiogenous cells (Fig. I.30). Are the conidia aleuriospores, sympodulospores, or phialospores (which is possible since there are phialides with solitary conidia (Gams, 1973) and since one often considers the first phialospore holoblastic)?

We thus group these fungi, if they are hyphal, in the Aleuriosporae, and if they are conidiomal, in the Monoblastosporae (the latter are found again in the Blastopycnidiineae and the monoblastic Blastostromatineae of Sutton, 1980).

Without wishing to multiply the examples and number of these problems, we feel that they demonstrate well the continuous nature of the types of conidiogenesis, which taxonomy artificially divides.* For our part, we have chosen to preserve some large groups for reasons of convenience (Tables I.H, I.I, I.J.), while trying to show that there are nevertheless some overlaps between them.

---

*At a more practical level, there are many other problems of choice for the taxonomist. There are, for example, intermediate forms between hyaline and melanized fungi, which leads one to disregard the division made by Saccardo between Mucedinaceae and Dematiaceae. Many of the terms of qualitative description are rather vague, such as broad/narrow, elongated/globular.

Table I.H. Types of conidiogenesis and the taxonomy adopted

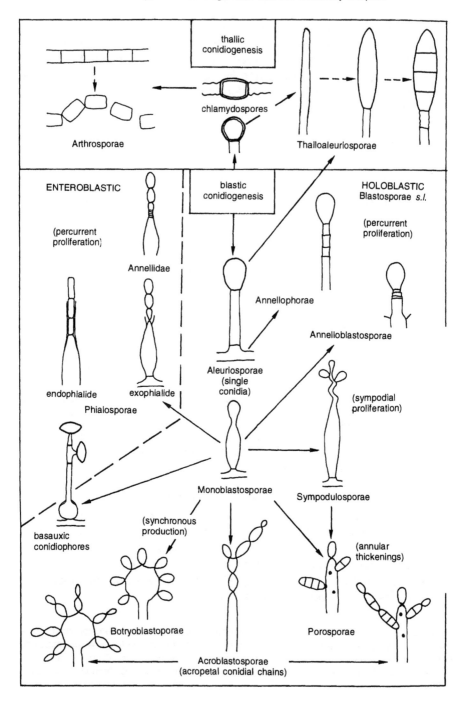

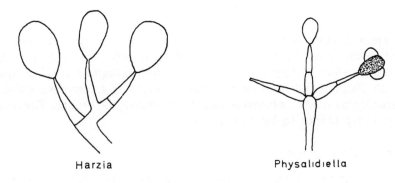

Harzia   Physalidiella

Fig. I.30. Examples of conidiogenesis that is difficult to characterize.

Table I.I. Grouping of conidiophores and the taxonomy adopted

Free conidiophores
HYPHALES
Grouped conidiophores
CONIDIOMALES
Conidiomata = coremia, sporodochia, acervuli, pycnidia, pseudopycnidia, thyriopycnidia, cupules

Table I.J. Types of conidiogenesis and the relevant chapters (see Figs. I.17-29 and Table I.H)

THALLOCONIDIA (arising from a pre-existing part of thallus)	
Arthrosporae	
Meristem Arthrosporae	ch. II
TRUE CONIDIA (arising from a process of budding)	
Holoblastic conidia = Blastosporae *s.l.*	ch. III
including	
Aleuriosporae	ch. IV
Annellophorae	ch. V
Sympodulosporae	ch. VI
Acroblastosporae	ch. VII
Botryoblastosporae	ch. VIII
Porosporae	ch. IX
Enteroblastic conidia	
Phialosporae	ch. X
Annellidae	ch. XI
Group with basauxic conidiophores	ch. XII
MYCELIA STERILIA	ch. XIII

Bibliography

The present work is devoted to the identification of genera, and it is therefore important to give a bibliography that directs the reader toward the identification of species. A more focussed bibliography will be given in each chapter, but here we list general works of references, either of essentially taxonomic interest or devoted to ecological groups. **These are the primary works to be consulted.**

Taxonomic groups

Dictionary of the Fungi by Hawksworth et al., 1995, covers numerous areas, notably:

— all the generic names, legitimate or otherwise, described up to the time of publication, and often a bibliographic lead that enables species identification;

— definition of mycological terms;

— guidelines on the usual laboratory techniques and the general biology of fungi.

Knowledge of the taxonomic groups can be updated from numerous publications, of which the following are essentially devoted to taxonomy:

— *Mycotaxon* (international review published in the USA);

— *Mycological Papers* and *Studies in Mycology*, specialized monographs published respectively by the CMI at Kew (UK) and by the CBS at Baarn (Netherlands);

— the series of descriptive cards published by the Institute of Biosystematic Researches of Ottawa (*Fungi Canadenses*) and by the *CMI* of Kew (*CMI Description of Pathogenic Fungi and Bacteria*);

— finally, one can also consult information databases such as the CAB Abstracts.

One may also consult the following contributions specifically on:

— Coelomycetes (according to the old classification, see Table I.B): von Arx, 1970; Grove, 1935, 1937; Melnik, 1997; Nag Raj, 1993; Punithalingam, 1970, 1974, 1981; Sutton, 1980.

— Hyphomycetes (see Table I.B): Braun, 1995; Carmichael et al., 1980; Ellis, 1971a, 1976; Matsushima, 1975; Subramanian, 1971; Sutton, 1973b.

— Deuteromycetes (belonging to the two preceding groups): von Arx, 1981; Sivanesan, 1984.

Ecological groups

The following publications are dedicated to fungi:

— of the air: Gregory, 1973; Nilsson, 1983; Wilken-Jensen and Gravesen, 1984.

— of food: Botton et al., 1990; Jesenska, 1993; Moreau, 1974; Pitt and Hocking, 1997.

— aquatic: Ingold, 1975.

— of wood: Mangenot, 1952, Wang and Zabel, 1990.

— coprophilic: Bell, 1983; Richardson and Watling, 1997.

— entomophagous: Samson et al., 1988

— fungicolous: Deighton and Pirozynski, 1972; Nicot, 1966.

— of grains and seeds: Malone and Muskett, 1997; Neergard, 1978.

— lichenicolous: Clauzade et al., 1989; Hawksworth, 1981.

— nematophagous: Barron, 1977; Ilyaletdinov and Pryadko, 1990.

— parasites of humans and animals: Badillet, 1991; Chabasse, 1988; Coudert, 1955; de Hoog and Guarro, 1996; Percebois, 1973; Segretain et al., 1979; Vanbreuseghem, 1966.

— parasites of vascular plants: Brandenburger, 1985; Ellis and Ellis, 1985; Funk, 1981, 1985; Messiaen et al., 1991; Morelet, 1978; Raynal et al., 1989; Smith et al., 1988.

— of the soil: Barron, 1968; Domsch et al., 1980; Singleton et al., 1993.

Methods and techniques of the study of fungi

The study of fungi necessitates a microscopic examination that may require a number of techniques: sampling, fine microsections, preparation in an appropriate mounting medium, microscopic observation, biometry, possibly microphotography and drawing, and dynamic study of the conidial ontogeny in culture on the slide.

Generally, the fungi are directly accessible on the substrates studied, but often it is necessary to isolate and culture them.

Useful guidelines on these techniques can be found in various works, such as those of Booth, 1971; Gams et al., 1998; Hennebert, 1977; Langeron and Vanbreuseghem, 1952; Malloch, 1981; Rapilly, 1968; Wastiaux, 1994.

In critical cases, it is necessary to use a scanning electron microscope and especially a transmission electron microscope for the characters of the cell wall and septa, or the order of conidiogenous events (Cole and Samson, 1979).

With the development of molecular techniques, above all in relation to the possibility of working with minute quantities of DNA, with the help of amplification (PCR), the field of application of molecular tools is widely increasing in fungal research. These techniques prove to be, in fact, extremely valuable in resolving, especially at the species and intraspecies level, problems of identification that could not be solved with morphological techniques. The generic level, which concerns us, poses in principle fewer problems of identification for a practised eye. All the same, in case of uncertainty, here also one can use ribosomal RNA analysis (especially 18 S and 26 S).

An abundant literature is already devoted to this field, and we can cite for example the works of Loncle et al., 1993; Bridge et al., 1998; Creighton, 1999; Innis et al., 1990.

Presently we have more or less numerous and detailed data about the molecular taxonomy of around 90 genera of Deuteromycetes. The existence of works of molecular identification for a given genus is indicated later in the last column, *Mol. Biol.*, of the legends in the generic tables, but the references are not cited there, for want of space. The interested reader can find them on the Web, from data banks such as: Genbank (www.ncbi.nlm.nih.gov/Web/Genbank/index.html) or EMBL-EBI (www.ebi.ac.uk/ebi-home.html).

Let us also recall that, as in molecular biology, the analysis of numerous biochemical characters can assist the morphologist studying taxa that are difficult to identify: immunological techniques (ELISA), physiological tests (vitamin needs, nitrogen and carbon assimilation, resistance to toxins, pH, temperatures, etc.), and chemotaxonomy (study of isoenzymes. polymers of the cellular wall, mycotoxins, sterols and fatty acids, ubiquinone systems, pigments, etc.).

ARTHROSPORAE AND MERISTEM ARTHROSPORAE

Arthrosporae

The term 'arthrospore', proposed by Vuillemin (1910b), designates a type of conidium that appears by fragmentation of a preexisting hypha (Fig. II.1). These are **thalloconidia**, as opposed to **true conidia**, which are produced *de novo* from a specialized conidiogenous cell.

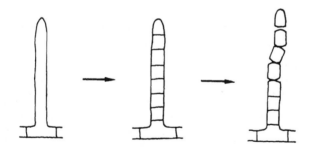

Fig. II.1. Fragmentation of the conidiogenous hypha into arthroconidia.

Generally, septation takes place only after the growth of the conidiogenous cell has stopped; the septa may be acropetal, basipetal, or random.

Tubaki (1963) distinguished exogenous (Fig. II.2a) and endogenous formation (Fig. II.2b) of arthroconidia. Kendrick (1971b) recognized six different types.

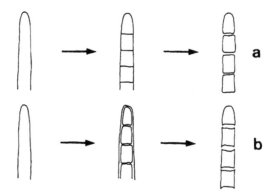

Fig. II.2. The supposed difference between exogenous arthroconidium (**a**) and endogenous arthroconidium (**b**).

For our part, by applying to some examples studied in electron microscopy, the examination of three principal stages—initiation of primordium, conidial maturation, and conidial secession—we distinguished three main types. The initiation of the primordium being fundamentally similar in the three types (i.e., reduced to the septation of the conidiogenous hypha), it is the maturation (retraction of cytoplasm in the hypha, bulging of conidia, modifications of the walls...) and secession (schizolytic or rhexolytic) that are different.

Some ultrastructural aspects of conidiogenesis

The process of arthroconidiogenesis has been the subject of detailed studies, which we will review here (Fig. II.3).

In the three types mentioned above, the initiation of primordium is similar: the septation of the conidiogenous hypha after the end of hyphal growth. The order of appearance of the septa may vary, however, and so may the details of septation: complete septa with plasmodesmata (pld) in *Geotrichum* (Fig. II.4), with a central pore in *Stephanosporium* and *Sporendonema*, these three genera being anamorphs of Ascomycetes; in the Basidiomycete *Pleurotus cystidiosus* (anamorph: *Antromycopsis broussonetiae*), the central dolipore becomes scarred and mixed up with the thickened wall of the spore. The zone of cleavage seems to function as in the Ascomycetes (Moore, 1977).

In the type *Geotrichum* (Fig. II.3) (Cole, 1975), each part defined by a septum becomes a conidium. Each septum is composed, according to the terminology of Reisinger (1972), of two layers B' in continuity with layer B of the wall of the parent hypha, on each side of the zone of cleavage (zc) (Fig. II.4).

Such a septum is symmetrical because the layers B' are similar on either side of the zone of cleavage. It is schizogenous, that is, the separation of conidia takes place at that point, by the rupture of the outer layers of the wall (A + one part of B), then cleavage of the median zone (zc). The extremities of the conidia become convex under

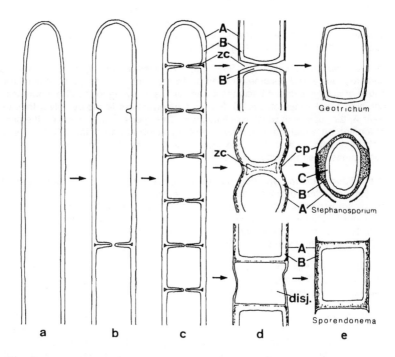

Fig. II.3. Diagram summarizing the ultrastructural studies of arthroconidiogenesis (according to Cole, 1975 and Reisinger, 1972).
a, b, c: initial process of septation common to all three types. In *Geotrichum* and *Stephanosporium*, **d:** all the cells of the filament become conidia and separate at the zone of cleavage, **zc**; in *Stephanosporium*, this becomes larger by autolysis. In *Sporendonema*, **d:** one cell out of two becomes a conidium, and the other degenerates to form a disjunctor, **disj**. Types of separation in **d** and **e:** typical schizolysis in *Geotrichum*, schizolysis with parietal connective in *Stephanosporium*, rhexolysis with cellular disjunctor in *Sporendonema*. **A, B, B', C:** different layers of the wall; **cp:** parietal connective; **disj.:** cellular disjunctor; **zc:** zone of cleavage.

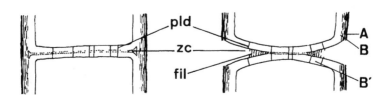

Fig. II.4. Schizolytic separation in *Geotrichum* (according to Cole, 1975).
Between two conidia the extremities become convex, enabling their separation at the zone of cleavage, **zc**.
A, B, B': different layers of the wall; **fil:** filaments of parietal material; **pld:** plasmodesmata; **zc:** zone of cleavage.

the internal turgor pressure of the cell (Fig. II.4); this is probably the mechanical cause of the centripetal schizogenous rupture of the septum. Soon the conidia are attached only by filaments that are visible under electron microscope. The bulging of the extremities of the conidia seems to cause an elongation and breaking up of the chain (Cole and Kendrick, 1969b) (Fig. II.1).

In the *Stephanosporium* type (Fig. II.3) (*S. cerealis*, Reisinger, 1972), the conidial primordia bulge and become rounded, while the zone of cleavage becomes larger because of autolysis until it becomes an empty space, owing to the disappearance of the non-pigmented zones of layer B. Finally, the conidia are linked by only a thin, residual layer A, the rupture of which releases them. This species has, moreover, the peculiarity of forming an equatorial ring made up of a melanized part of layer B; the conidia are diplothecate because they have a layer C. In *Oidiodendron griseum* (Terracina, 1977), sporogenesis seems quite comparable.

In the *Sporendonema* type (Fig. II.3) (*S. purpurascens*, Cole, 1975), after segmentation, all the cells thus delimited do not become conidia: in around one out of two, the cytoplasm becomes condensed and the wall thickens, while the others degenerate. We therefore finally have conidia with dense contents and relatively thick walls alternating with disjunctors that have more or less broken-up cytoplasm and thin walls (alternate arthroconidia, Sigler and Carmichael, 1976).

The septa in this case are asymmetrical (thicker near the conidium) and not schizogenous. The separation occurs by the rupture of the thin residual wall of the separating cells and is therefore rhexolytic.

The first type, seen in *Geotrichum*, is considered exogenous = holoarthric (all the layers of the parent hypha are involved in the formation of a conidial wall). *Sporendonema*, on the contrary, represents an endogenous type = enteroarthric, in which the 'primary wall' is not part of the conidial wall: it is considered after observations under light and scanning electron microscope that the conidia form on the **inside** of this 'primary wall', indeed inside the hypha, and they are able to emerge from it (Cole, 1975; Sigler and Carmichael, 1976). Our diagram, based on unique photographs taken under transmission electron microscope (Cole, 1975), shows that all the layers of the wall of the parent hypha are found in that of the conidium, which is no more endogenous than in *Geotrichum*. The TEM studies of arthrospores are too rare to support general conclusions; however, the difference does not seem to be between holoarthric (exogenous) conidia and enteroarthric (endogenous) conidia (Terracina, 1977; Martinez et al., 1986), but rather between conidia of schizolytic and rhexolytic separation. There does not seem to be an intermediate between these two types. Although, in *Stephanosporium*, the distended layer A has the appearance of a disjunctor (this is what we call a **parietal connective**), this space results from the transformation of **a single septum** that is schizogenous. In *Sporendonema*, the disjunctor is a degenerated cell delimited by **two septa** that are non-schizogenous.

Conidiophores

The conidiophores are quite variable in this group. In *Geotrichum*, they are nonexistent or micronematous because they are mixed with the mycelium. In the genus *Oidodendron*, they are macronematous, raised, and easily differentiated. *Sympodiella* has the peculiarity of producing conidiogenous hyphae by sympodial, subapical growth. These hyphae break up into conidia, then, after the secession of the latter, the conidiophore appears as a rachis with alternating scars, recalling certain Sympodulosporae.

Conidia

Conidia formed by the process of septation of the thallus, explained above, can only slightly increase in volume after they are enclosed. They can become rounded and often globular, barrel-shaped, or cylindrical, small and unicellular.

Some *Chrysosporium* form intercalary arthroconidia, in addition to characteristic terminal aleurioconidia. The relationships between chlamydospores and arthroconidia are particularly clear-cut in *Chrysosporium* and *Scytalidium*, in which all the intermediates between these two types of spores are found.

Sometimes fungi of another type are confused with Arthrosporae:
— Meristem Arthrosporae (e.g., *Basipetospora*, Fig. II.6), which we will examine; the distinction is not always clear.
— Deuteromycetes among which the conidia, of different ontogeny, break up at maturity (e.g., the aleuriospores of *Chalara elegans*, Fig. II.5a; the sympodulospores of *Wiesneriomyces*, Fig. II.5b). Minter and Holubova-Jechova (1981) named these conidia secondary arthrospores, in *Blastophorum*.

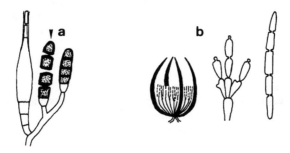

Fig. II.5. a: Breaking up (arrow) of an aleuriospore of *Chalara elegans*; **b:** divisible sympodulospore of *Wiesneriomyces*, mature spore at right.

Key to identification of Arthrosporae

1	Conidiomata absent	2
	Conidiomata present	22

Hyphales _____

2(1)	Undifferentiated conidiophores	3
	Conidiophores more or less differentiated	15
3(2)	Schizolytic secession	4
	Rhexolytic secession	10
4(3)	Arthroconidia hyaline	5
	Arthroconidia brown	8
5(4)	Only arthroconidia	6
	Arthroconidia with other spores	7

6(5)	Arthroconidia mucous	*Geotrichum*	**A13***
	Arthroconidia dry	*Mauginiella*	**A13**
7(5)	Arthroconidia + brown chlamydospores	*Scytalidium*	**A15**
	Arthroconidia + blastoconidia and sometimes chlamydospores	*Moniliella*	**A12**
8(4)	Only arthroconidia		9
	Arthroconidia + brown chlamydospores	*Arthropsis*	**A14**
9(8)	Uni- or multicellular	*Bahusakala*	**A19**
	Regularly bicellular, presence of hyphopoda	*Ampulliferina*	**A20**
10(3)	Hyaline conidia		11
	Melanized conidia, cellular disjunctors	*Arthrocristula*	**A5**
11(10)	Conidia more or less cylindrical, in long chains		12
	Conidia ovoid		13
12(11)	Diameter less than 4 μm	*Malbranchea*	**A2**
	Diameter more than 4 μm, cellular disjunctors	*Sporendonema*	**A1**
13(11)	Parent hypha with determined growth		14
	Parent hypha with undetermined growth	*Arcuadendron*	**A7**
14(13)	Conidia in long chains	*Ovadendron*	**A8**
	Conidia in short chains	*Chrysosporium*	**A9**
15(2)	Schizolytic secession		16
	Rhexolytic secession		20
16(15)	Schizogenous septa		17
	Parietal connectives		19
17(16)	Conidiophore with sympodial growth	*Sympodiella*	**A16**
	Conidiophore with non-sympodial growth		18
18(17)	Conidiophores with verticils of conidiogenous cells (synanamorph *Malbranchea*)	*Oncocladium*	**A17**
	Conidiogenous cells on lateral branches	*Arthrographis* (=*Oidiodendron*?)	**A18**
19(16)	Annular thickenings around the conidia	*Stephanosporium*	**A3**
	No such character	*Oidiodendron*	**A6**
20(15)	Conidiophores poorly differentiated, short chains of conidia	*Chrysosporium*	**A9**
	Conidiophores well differentiated, long chains		21
21(20)	Conidiophores branched at an acute angle	*Geomyces*	**A10**
	Apical verticil of fertile branches	*Amblyosporium*	**A11**

Conidiomales

22(1)	Coremia	23
	Other types of conidiomata	25

---

*See also *Trichosporonoides* (chapter VIII), ·in which the hyphae are segmented into arthrospores.

23(22)	Schizolytic secession		24
	Rhexolytic secession	*Coremiella*	**B2**
24(23)	Conidia in dry chains	*Briosia*	**B3**
	Conidia in slimy heads (monokaryotic small and ovoid, dikaryotic larger, with an outgrowth arising from clamp connexions) (see also the related genus *Tilachlidiopsis* X E3)	*Antromycopsis*	**B1**
25(22)	Schizolytic secession		26
	Rhexolytic secession		36
26(25)	Sporodochia or acervuli		27
	Other conidiomata		30
27(26)	Conidia hyaline		28
	Conidia brown		29
28(27)	Sporodochia	*Cylindrocolla*	**B7**
	Acervuli	*Septotis*	**B9**
29(27)	Conidia most often bicellular, with mucilaginous sheath	*Staninwardia*	**B10**
	Multiseptate conidia without mucilaginous sheath	*Septotrullula*	**B8**
30(26)	Cupuliform conidiomata at least at maturity		31
	Pycnidia	*Neozythia*	**B16**
31(30)	Superficial cupules, ornamented with bristles at the edge	*Xiambola*	**B11**
	Submerged pseudopycnidia, opening into cupule at maturity		32
32(31)	Highly developed basal stroma, rounded conidia	*Ojibwaya*	**B4**
	Undeveloped basal stroma		33
33(32)	Conidia with cellular appendages	*Acarosporium*	**B15**
	Conidia lacking appendages		34
34(33)	Conidia uni- to multicellular	*Sirozythiella*	**B14**
	Conidia unicellular		35
35(34)	Conidia hyaline, elongated	*Phacidiella*	**B13**
	Conidia slightly coloured, shorter	*Trullula*	**B12**
36(25)	Conidia hyaline, with clamp connexions	*Ptychogaster*	**B5**
	Conidia pale brown	*Geotrichella*	**B6**

Meristem Arthrosporae

In these fungi, the conidia in basipetal chains result from a 'retrogressive' evolution of the conidiophore:

In the first type (Fig. II.6), the length of the sporogenous part is fixed from the beginning. The tip is transformed into conidia from the top towards the base, the length of the conidiophore becoming smaller and smaller each time.

Table II.A. Hyphal Arthrosporae

Fig.	Genus	No. spp.	Geographical distribution	Mode of life, substrates and impact	Teleomorphs	Additional references	Mol. biol.*
1	*Sporendonema*	3	Cosmopolitan	Caseicolous, fungicolous			
2	*Malbranchea*	17	Cosmopolitan	Telluric, keratinophilic or cellulolytic	*Auxarthron Myxotrichum Uncinocarpus*		X
3	*Stephanosporium*	1	Cosmopolitan	Cellulolytic, lignolytic, mostly soil and roots			
4	Anamorphs of Basidiomycetes**						
5	*Arthrocristula*	1	USA, Sri Lanka	Telluric			
6	*Oidiodendron*	18	Cosmopolitan	Cellulolytic or keratinolytic, mostly plant litter and debris	*Byssoascus Myxotrichum*	Mycot. 28: 233 (1987)	X
7	*Arcuadendron*	2	Eurasia	Telluric, keratinophilic			
8	*Ovadendron*	1	Europe	Keratinolytic, parasite on humans			
9	*Chrysosporium*	40	Cosmopolitan	Telluric, keratinophilic, cellulolytic and/or pathogen of humans and animals, coprophilic, glucophilic	*Aphanoascus Arthroderma Gymnoascus* and other Onygenales	Stud. Mycol. 20 (1980)	X
10	*Geomyces*	4	Ubiquitous, above all the temperate North	Telluric	*Pseudogymnoascus*	Stud. Mycol. 20 (1980)	
11	*Amblyosporium*	3	Temperate North	Telluric, fongicolous			
12	*Moniliella*	2	Northern hemisphere	Food products, tobacco		Stud. Mycol. 19 (1979)	
13a	*Geotrichum*	23	America, Europe	Telluric, mycoparasite industrial and food contaminant, agent of geotrichosis	*Dipodascus Galactomyces*	Stud. Mycol. 29 (1986)	X
13b	*Mauginiella*	1	North Africa, Middle East	Flower rot of *Phoenix dactylifera*			
14	*Arthropsis*	4	Peru	Dead leaves		Mycot. 54: 281 (1995)	
15	*Scytalidium*	15	Cosmopolitan	Telluric, fungal antagonist, hyperparasite of Uredinales, agent of dermato- and onychomycosis, keratinolytic, lignicolous, mycorrhizian (1 sp. on very acid mineral media, 1 sp. agent of composting of straw)	*Hymenoscyphus*	Mycot. 25: 279 (1986)	X
16	*Sympodiella*	2	Europe, Cuba	Foliicolous			
17	*Oncocladium*	1	America, Europe	Telluric, keratinolytic	*Gymnoascus*		
18	*Arthrographis*	5	Cosmopolitan	Keratinolytic, cellulolytic, lignolytic	*Pithoascina*		
19	*Bahusakala*	5	Ubiquitous	Saprophyte or weak parasite on stem (*Pinus, Salix, Yucca*) termite nests	*Aulographina*		
20	*Ampulliferina*	1	America, Europe	Dead leaves (*Fagus, Ledum*)			

*Mol. biol. = Molecular biology. The presence of an X in this column indicates the existence of a study of at least one species by this technique.
**Several Basidiomycetes have this type of anamorph, which has not been named.

Hyphal Arthrosporae

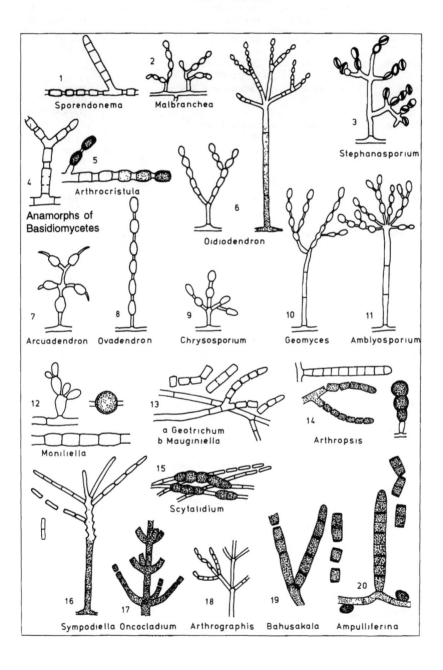

Table II.B. Arthrosporae with conidiomata

Fig.	Genus	No. spp.	Geographical distribution	Mode of life, substrates and impact	Teleomorphs	Additional references	Mol. biol.
1	Antromycopsis	3	Ubiquitous	Saprophyte, lignicolous	Pleurotus (B)	CJB 69: 6, 1991	X
2	Coremiella	1	Europe, USA	On various dead or dying plant organs			
3	Briosia	2	America, Europe	Pathogenic on Vitis leaf		Mycot. 4: 349 (1976)	
4	Ojibwaya	1	Canada, Malawi	Saprophyte on Juniperus stem			
5	Ptychogaster	10	Ubiquitous	Lignolytic	Tyromyces (B) Oligoporus (B)	Mycot. 4: 349 (1976)	
6	Geotrichella	2	Alaska, France	On paper		Mycot. 4: 349 (1976)	
7	Cylindrocolla	5	Temperate North	1 species frequent on dead stem of Urtica	Calloria, Calloriella		
8	Septotrullula	1	Europe	On bark (Betula, Quercus, Fagus)		TBMS 48: 355 (1965)	
9	Septotis	2	North America, Europe	Pathogenic on leaves (Populus, Podophyllum)	Septotinia	Mycol. 72: 208 (1980)	
10	Staninwardia	1	Southeast Asia	Pathogenic on leaves of Eucalyptus			
11	Xiambola	1	Czechoslovakia	On decomposed needles of Pinus		Fol. G. Pr. 16: 195 (1981)	
12	Trullula	15	Temperate North	1 species frequent on cones of Abies			
13	Phacidiella	3	Eurasia	On dead twigs (Salix) or dead stems (Artemisia, Asperula)	Pyrenopeziza		
14	Sirozythiella	1	Europe	On leaves and culms (Phragmites)			
15	Acarosporium	2	North America, Europe	On various organs of trees (Alnus, Betula, Acer, Populus, Salix)	Pycnopeziza		
16	Neozythia	1	Kurdistan	On dead stem of Astragalus			

In the Arthrosporae, we have seen that there is no growth of the conidiophore after the production of conidia. In the other Deuteromycetes that have **blastic** conidiogenesis, there is generally an **apical** growth (for our purposes we can combine the growth with **wall production**). Here a **diffuse** growth (parietogenesis) happens, that prevents the growth of the tip towards the top, causing it to bulge and become rounded before sep-

Arthrosporae with conidiomata

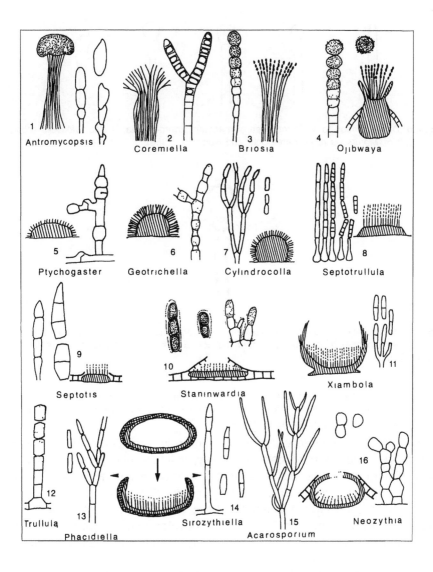

tating to form a new conidium; this type is thus intermediate between thalloconidia and true conidia (Fig. II.7). In this species, the process ends with one single conidiogenous cell.

In *Basipetospora variabilis*, the conidiophore is branched and septate: when the conidiogenous terminal cells are entirely converted, the process extends to the lower cells and ends with the formation of ramified chains of conidia. In *B. chlamydosporis*, things

Fig. II.6. Differentiation towards the base of meristem arthroconidia, in *Basipetospora rubra*, anamorph of *Monascus ruber* (after Cole and Kendrick, 1968).

Fig. II.7. Detail of the formation of a conidium in *B. rubra*. The apical part of the conidiophore (wall shown in white) bulges (arrow) under the conidium already formed, to differentiate a new conidium.

occur at first as in *B. variabilis*, but after production of the last conidium, according to Minter et al. (1983a), there is a recurrence of growth of the conidiophore by an **annular** (intercalary) parietogenesis as shown in Fig. II.9.

Similarly, in *Trichothecium roseum*, there is formation of new conidia under the conidium produced previously, with correlating shortening of the conidiogenous cell, but here an apical growth (lateral in fact) takes place, ending in a chain of alternating conidia (Fig. II.8).

In the second type, the length of the conidiophore is not fixed: the conidia, often phragmo- or dictyospores, develop and mature from the tip to the base of the chain, while the conidiophore continues to grow by annular parietogenesis.

This type is found in the genera *Coniosporium, Phragmotrichum, Trimmatostroma*, and *Barnettella*, as well as in the anamorphs of Erysiphaceae (Zheng, 1985).

According to Minter et al. (1983a), *Basipetospora variabilis* has a conidiogenesis differing little from that of an arthrospore of *Oidiodendron*, for example: to put it simply, the individualization is strictly basipetal in the former, instead of being more or less random, as in the latter, and the conidia become rounded before being closed, not after.

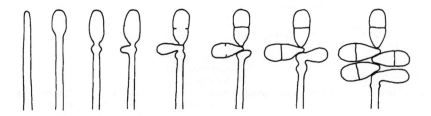

Fig. II.8. Lateral and retrogressive formation of conidia in *Trichothecium roseum* (after Kendrick and Cole, 1969).

Fig. II.9. Retrogressive conidiogenesis with growth of conidiophore, by annular parietogenesis (arrows).

In *B. chlamydosporis*, the annular growth is not continuous, as in the meristem Arthrosporae of the second type, but sequential: these are two examples of intermediates between types of conidiogenesis. Some sequences of events are reversed or modified, but the phenomena are not fundamentally different.

We have already mentioned in the introduction the existence of a continuum between fungal species, poorly understood by means of artificial divisions of taxonomy.

To the meristem Arthrosporae we add *Wallemia*, in which the sporogenous cell resembles a phialide, but it puts out a filament that then divides into arthrospores. A new filament may then be produced, pushing out the first series of arthrospores, then fragmenting itself, etc. After several cycles, a new 'phialide' may develop by percurrent growth, which itself produces new conidiogenous filaments. We must note that it is difficult to define the conidiogenous cell here: is it the 'phialide' or the filament that emerges from it (Hashmi and Morgan-Jones, 1973; Terracina,

Table II.C. Meristem Arthrosporae

Fig.	Genus	No. spp.	Geographical distribution	Mode of life, substrates and impact	Teleomorphs	Additional references	Mol. Biol.
1	*Wallemia*	1	Cosmopolitan	Osmophilous, on food, sometimes textiles, hay, household dust, humans	(Basidiomycete)		
2	*Oidium*	± 120	Cosmopolitan	Pathogen on many plants	*Erysiphe, Microsphaera, Podosphaera, Sphaerotheca, Uncinula*	Fl. Jena 170: 77 (1980)	X
3	*Basipetospora*	4	Ubiquitous	Telluric, amylolytic	*Monascus*		
4	*Trichothecium*	5	Cosmopolitan	Agent of fruit rot, saprophyte or weak parasite on various plant organs, toxinogenic	*Heleococcum, Hypomyces*		
5	*Vouauxiella*	4	South America, Europe	On lichens			
6	*Barnettella*	3	Indies	On Gramineae			
7	*Trimmatostroma*	9	Ubiquitous	Stems and branches (*Betula, Populus, Salix*), litter (*Pinus*)			
8	*Coniosporium*	2	North America, Europe	On dead wood	*Hysterium*		
9	*Phragmotrichum*	5	Eurasia, USA	Scales of cones of *Picea*, twigs of numerous deciduous or coniferous trees		TBMS 48: 349 (1965)	

Meristem Arthrosporae

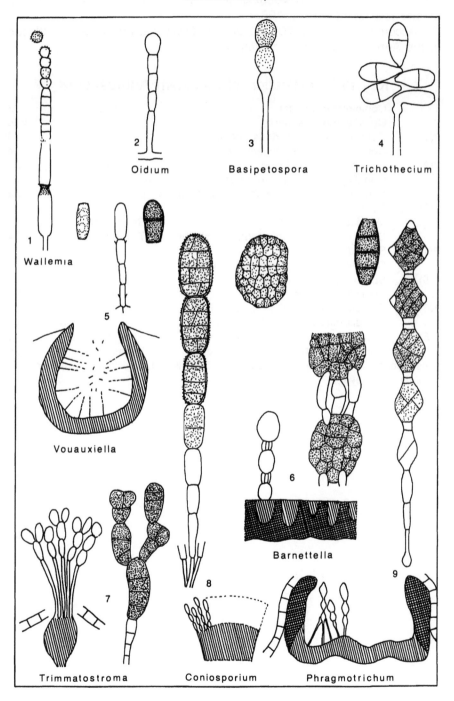

2
Oidium

3
Basipetospora

4
Trichothecium

1
Wallemia

5

Vouauxiella

6

Barnettella

7
Trimmatostroma

8
Coniosporium

9
Phragmotrichum

1974)? The conidiomale *Vouauxiella lichenicola* presents the same con-idiogenesis (Minter *et al.*, 1983a).

According to Ellis (1976), *Briosia* and *Ojibwaya*, which we have placed in the Arthrosporae, are meristematic.

Key to identification of meristem Arthrosporae

1	Conidiomata absent		2
	Conidiomata present		5
2(1)	Conidiogenous cell does not diminish		3
	Conidiogenous cell diminishes		4
3(2)	Appearance of phialide	*Wallemia*	C1
	Conidiogenous cell with intercalary growth, at the base of the conidial cell	*Oidium*	C2
4(2)	Conidia unicellular, appearance of aleuriospores	*Basipetospora*	C3
	Conidia bicellular, in alternating chains	*Trichothecium*	C4
5(1)	Sporodochia		6
	Conidiomata cupuliform or pycnidia		8
6(5)	Sporodochia entirely fertile		
	Sporodochia with fertile alveoli	*Barnettella*	C6
7(6)	Conidia multicellular in simple chains	*Coniosporium*	C8
	Conidia multicellular in branched chains	*Trimmatostroma*	C7
8(5)	Conidiomata cupuliform at maturity	*Phragmotrichum*	C9
	Conidiomata: in unilocular pycnidia	*Vouauxiella*	C5

(See also the Arthrosporae with basipetal differentiation, *Briosia, Antromycopsis, Trullula, Ojibwaya....*)

BLASTOSPORAE
sensu lato

As we have tried to show in the introduction, these groups have in common a typically **blastic** conidiogenesis, that is, the conidial primordium is produced by a process of budding (Fig. III.1).

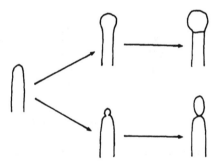

Fig. III.1. Blastic conidiogenesis with a wide base (above) and a narrow base (below).

Of course, conidiogenesis is also blastic in the Phialidae. But in these, the conidia are produced inside the neck of the conidiogenous cell, while the Blastosporae produce external spores (see holoblastic and enteroblastic conidiogenesis, Fig. I.6).

The initiation of the conidial primordium is very similar in the different groups (Fig. III.1), and it is the details of the other processes of conidiogenesis and development of the conidiophore that differ.

In the Aleuriosporae, the base of the conidium is in principle as large as the tip of the conidiogenous cell (Fig. III.2).

The Sympodulosporae, Acroblastosporae, Porosporae, and Botryosporae, in contrast, generally have a narrow vegetative point. The

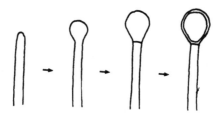

Fig. III.2. Conidiogenesis with a wide base of the aleuriospore.

Fig. III.3. Hilum and conidial scar in the case of conidiogenesis with a narrow base.

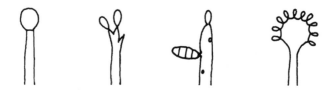

Fig. III.4. Examples of a single conidium per vegetative point.

hilum and the conidial scar may consequently be pointed, but that is not always the case (Fig. III.3).

In the Aleuriosporae, the Sympodulosporae, certain Porosporae, and certain Botryoblastosporae, each vegetative point has a single conidium (Fig. III.4).

In the Acroblastosporae, other Porosporae, and other Botryoblastosporae, each conidium can bud into one or several others (Fig. III.5).

The conidial chains are basifugal (= acropetal). The conidiophore may emerge in a sympodial fashion (Fig. III.6) or percurrently (Fig. III.7).

Each meristematic point located above the preceding one leaves a scar, which causes the ringed appearance of the conidiophore (= holoblastic Annellophorae, chapter V).

Since these two types of growth can coexist within a single genus, even in a single species, we have grouped together the fungi that

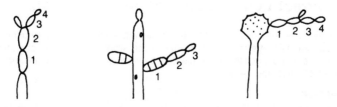

Fig. III.5. Examples of chains of successive conidia arising from a single vegetative point.

Fig. III.6. Sympodial proliferation of conidiophore.

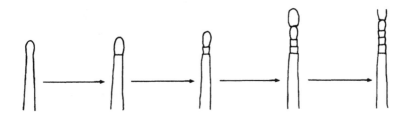

Fig. III.7. Percurrent proliferation of conidiophore.

present them under the large group of Blastosporae *sensu lato*. (It has been seen in the introduction that other Annellosporae are enteroblastic, chapter XI.)

We also include here the Monoblastosporae (above all the conidiomal ones), in which each conidiogenous cell produces only one conidium.

The principal characteristic of Blastosporae *sensu lato* seems to be that each meristematic point corresponds to a single conidium. Multiple conidia are produced by multiplication of meristematic points, synchronously (Botryoblastosporae, chapter VIII) or successively, and in this case by sympodial growth (Sympodulosporae, chapter VI, and certain Porosporae, chapter IX), or percurrent growth (holoblastic Annellosporae, chapter V). They can also occur by the production of acropetal conidial chains (Acroblastosporae, chapter VII, certain Porosporae and Botryoblastosporae).

Ultrastructural study of conidiogenesis

A sufficiently large number of Blastosporae *sensu lato* have been studied under electron microscope. We will attempt to summarize these works in applying our attention to conidiogenesis and especially to the relations between conidiogenous cell and conidium at the level of walls and septa.

Aleuriosporae

In all cases the initiation of conidial primordium is similar: apical budding with a wide base from the conidiogenous cell. The closure of the primordium then takes place, by appearance of a basal septum, and that sometimes very rapidly, well before the conidium has attained its definitive size. Maturation involves the production of a thick conidial wall, often melanized. The basal septum becomes asymmetrical, being thicker and melanized on the conidium side.

Separation occurs by rupture of the wall of the conidiogenous cell, and is therefore rhexolytic, in *Humicola grisea* (Fig. III.8).

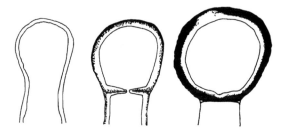

Fig. III.8. Formation of aleuriospores in *Humicola grisea* (after Griffiths, 1974).

Wardomyces pulvinatus generally releases conidia in a rhexolytic manner (cf. Fig. I.14), the propagule carrying all or part of the conidiogenous cell emptied of its contents: more rarely, separation is schizolytic (cf. Fig. I.14). The conidiogenous cell, the wall of which is melanized asymmetrically, may grow laterally from the less pigmented side to produce secondary conidia (Fig. III.9c).

Epicoccum nigrum presents a partly schizolytic secession, a phenomenon emphasized by the swelling of the basal cell and of the tip of the conidiophore (Fig. III.10). The semi-septa become separated towards the septal pore, then the internal pressure causes one of these two cells to burst, leaving a broken zone visible under light microscope and scanning electron microscope (Traquair and Kokko, 1981). This mechanism causes the active release of spores.

Holoblastic Annellosporae

Annellophorae

Acrogenospora sphaerocephala forms globular conidia at the tips of long, erect conidiophores: the first conidium is typically an aleuriospore, formed by the process described earlier. The secession is rhexolytic, but often the conidium does not detach itself immediately after maturation (Fig. III.11). It produces a regeneration of the tip by growth of a portion from the lower septum (Fig. III.11b), making its way in the conidiogenous cell like an intrahyphal hypha, which forms a conidium at a higher level, from which arises the ringed appearance of the conidiophore (Fig. III.11c). If the preceding conidium does not fall off, it is thrust aside (Fig. III.11b). Sometimes an aborted

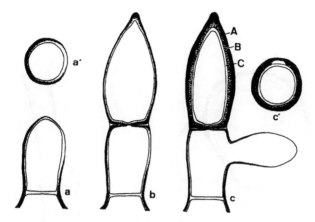

Fig. III.9. Formation of aleuriospores and lateral growth in *Wardomyces pulvinatus* (after Reisinger, 1970a, 1972).
a, a': longitudinal and cross-sections of the conidiogenous cell. Note the asymmetrical melanization of the wall; **b:** conidiogenous cell and immature conidium; **c:** mature conidium, with walls highly melanized and having a layer C. Formation of a young conidium by lateral growth from the less melanized side of the sporogenous cell; **c':** cross-section of the conidium with germ slit.
A, B, C: different layers of the cell wall.

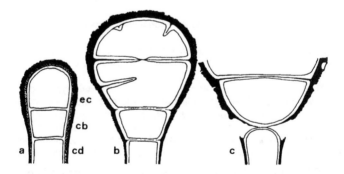

Fig. III.10. Formation of aleuriospores in *Epicoccum nigrum* (after Griffiths, 1973).
a: young conidial primordium (**ec**), basal cell (**cb**), and conidiophore (**cd**); **b:** growth and septation of the conidium; **c:** conidial separation that is partly schizolytic, by swelling of the basal cell and of the tip of the conidiophore, and separation at the zone of cleavage.

conidium is crossed by the growth of the conidiophore. In *Acrogenospora*, the semi-septum located under the conidium do not make up part of the new tip. Such are the facts observed in culture (Reisinger, 1972; Hammill, 1972b). However, Hughes (1979) opined that this mode of proliferation by intrahyphal hyphae is an anomaly due to the culture and that regeneration occurs in nature from the semi-septum remaining after schizogenous secession, in *Acrogenospora*. The process of regeneration by intrahyphal hyphae is, on the other hand, quite real in *Endophragmiella* and several related genera

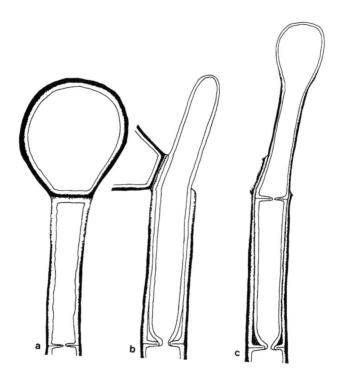

Fig. III.11. Formation of an annellophore in *Acrogenospora sphaerocephala* (after Reisinger, 1972; Hammill, 1972b).
a: first conidium formed; **b:** its partial secession and falling off, intrahyphal growth of the conidiogenous cell from the lower septum; **c:** formation of a new conidium.

(Table V.B). Anyway, the genus *Acrogenospora* is absolutely typical of the group of Annellophorae: no doubt, they behave like Aleuriosporae in which a percurrent proliferation of the conidiophore occurs. Each conidium has the same relations with its conidiogenous cell that an aleuriospore has with its own: initiation of primordium, maturation, and separation are very similar. More particularly, each conidium has a conidial wall continuous with that of its conidiogenous cell and remains unique.

Wang (1990) named this type of conidiophore an **annellophore**, characterized by intrahyphal growth and holoblastic conidia. It is entirely different and separate from the conidiogenous cell, which has percurrent proliferation but is otherwise similar to the phialide, the **annellide** with enteroblastic conidia (see chapter XI, Annellidae). However, the following type appears to be intermediate between the two.

Annelloblastosporae

Two species of this group that form conidiomata, *Melanconium bicolor* and *M. apiocarpum*, have been studied by Sutton and Sandhu (1969) (Fig. III.12).

The first conidium is, as always, holoblastic (Fig. III.12a). The cytoplasm of the conidiogenous cell is retracted more or less quickly, after formation of the conidium, while

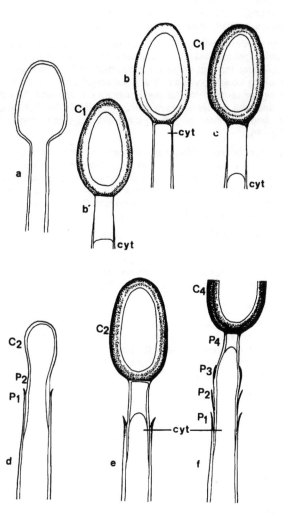

Fig. III.12. Formation of an annellophore in *Melanconium* spp. (after Sutton and Sandhu, 1969).
a, b, b', c: production of the first conidium; **b':** precocious retraction of cytoplasm of conidiogenous cell; **b, c:** late retraction with formation of a lower semi-septum; **d, e, f:** successive growths of a conidiophore forming an annellophore.
C1, C2,...: successive conidia; **P1, P2,...:** successive scars of the wall; **cyt:** cytoplasm.

its basal septum appears. Depending on the case, this septum can be made up of just one layer, the upper semi-septum formed by the cytoplasm of the conidium itself (precocious retraction, Fig. III.12b') or of the two semi-septa, elsewhere of unequal thickness, the lower semi-septum formed by the conidiogenous cell being thin and slightly melanized (later retraction, Fig. III.12b, c). The cytoplasm of the conidiogenous cell seems to retract after the production of each conidium, on a relatively constant length of 3-5 μm. It thus leaves an upper part of the conidiogenous cell empty, with a thin and slightly melanized wall and, although the photographs of Sutton and Sandhu do not show it clearly, it seems evident that the secession is rhexolytic and occurs according to the mechanism shown in Fig. I.14b or b'.

The cytoplasm then grows to produce a new wall P2, internal in relation to the primary wall of the conidiogenous cell P1. A conidium C2 will be produced, at a higher point. There will be, as for the first conidium, production of a transverse septum, retraction of cytoplasm of the conidiogenous cell, and secession of the conidium. This separation taking place higher up than for the first conidium C1, the wall P2 will therefore stick out over P1, and their border will form a ring (Fig. III.12d, e). The process is repeated and the superimposed rings give the parent cell a typically ringed (annellated) appearance (Fig.12f).

In relation to *Acrogenospora*, *Melanconium* seems to differ principally in that the growth does not arise from a septum (lower septum of the conidiogenous cell in the case of rhexolytic secession, higher in the case of schizolytic secession), but only from the point where the cytoplasm of the conidiogenous cell is retracted. According to what can be observed in the micrographs of Sutton and Sandhu, the new wall is produced from this point of retraction of cytoplasm (Fig. III.12d, e).

In relation to the Annellidae, which are studied in chapter XI, the retraction of the cytoplasm of the conidiogenous cell and the rhexolytic secession observed in *Melanconium* seem to us to be essential differences. But they are visible only under electron microscope and, as the resemblances between Annelloblastosporae and Annellidae are numerous, it is difficult to distinguish between the two groups.

Sympodulosporae

In the anamorph of *Poronia punctata* (Ascomycete, Xylariaceae), the conidia are produced laterally on the hyphae by the intermediary of little denticles. Fig. III.13 depicts the principal steps of conidiogenesis. It shows that all the layers of the wall of the conidiogenous cell are involved in the development of that of the conidium and that the secession is schizogenous.

Acroblastosporae

Cladosporium herbarum: the observations of Roquebert (1981) show here again that the fundamental layer B of the conidial wall is continuous with that of the conidiogenous cell, whether this be a conidiophore or a conidium. In fact, on the one hand the

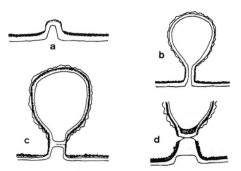

Fig. III.13. *Poronia punctata*, anamorph, according to Stiers *et al.* (1973).
a: lateral protrusion of the conidiogenous hypha; **b, c:** growth, maturation and closure of the conidium; **d:** schizogenous dehiscence.

conidiophore can grow subapically (Fig. III.14d), and on the other hand a conidium can bud into others (Fig. III.14e). Secession is schizolytic.

Alysidium resinae

This fungus, similar to the preceding species, differs from it by the thickness and often strong melanization of its walls, the width of its conidial insertions, and above all the characteristics of its conidiogenesis, which has been considered enteroblastic, which would be an exception among the Blastosporae s.l. We feel in fact that this interpretation does not correspond to reality:

— There is rupture, at the initiation of the conidial primordium, of the melanized outer layers A and B2 of the wall of the conidiogenous cell (Fig. III.15a).

— The primordium and then the conidium are continuous with the inner layer B1 of the conidiogenous cell (Fig. III.15a, b, c). Since B1 is the young, 'living' part of the wall and is actually involved in the formation of the conidium, the walls of the conidiogenous cell and the conidium are continuous and the conidiogenesis is holoblastic.

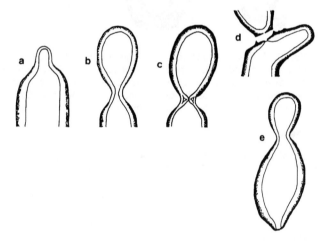

Fig. III.14. *Cladosporium herbarum* (after Roquebert, 1981).
a-c: steps in the formation of a conidium, apical bulging of the conidiogenous cell up to the basal septation; **d:** secession of a conidium and subapical growth of the conidiophore; **e:** production of a secondary conidium.

Subsequently, layer B of the conidium becomes melanized on the outer side, forming a layer B2 and a layer A (Fig. III.15d, e), while a layer C appears on the inner side: the conidia of *Alysidium* are therefore diplothecate (Fig. III.15f).

Torula spp. (sensu stricto: T. herbarum, T. caligans, T. terrestris)

Ellis and Griffiths (1975a and b) studied these three species, which are representatives of the genus in its present sense (= Hyphales, phragmosporate dematiaceous, in ramified and basifugal conidial chains). But while Hughes (1953) places it in section VI (the Porosporae), Subramanian (1962) designates it the type of family Torulaceae corresponding partly to Blastosporae s.l.

The electron microscope shows that the conidia of *Torula* are well-differentiated acroblastospores, with median cells having a layer C, thus distoseptate, a specialized, apical conidiogenous cell, and sometimes a differentiated basal cell (Fig. III.16a).

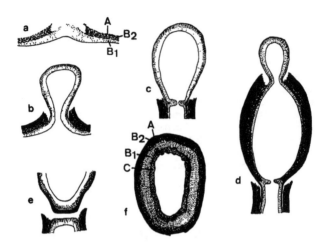

Fig. III.15. *Alysidium resinae* (after Ellis and Griffiths, 1977).
a, b: formation of a conidial primordium; note the rupture of layer B2 of the conidiogenous cell; **c:** appearance of a septum at the base of the young conidium; **d:** maturation of the conidium, with melanization leading to the differentiation of a new layer B2, and formation of a secondary conidium; **e:** conidial secession and details of parietal layers; **f:** mature conidium.
A, B1, B2, C: different layers of the wall.

The structure of the wall is distinctive in the conidiogenous cell, lacking a layer C, melanized in the basal part, hyaline at the tip. At maturity, it produces a differentiation at the base of the conidium and at the tip of the conidiogenous cell: melanized layer B2, which surrounds the basal cell of the upper conidium, is extended to a cap in the hyaline tip of the lower conidiogenous cell (Fig. III.16c). This structure is different from the typical pore of the Porosporae, which we will look at later.

Botryoblastosporae

Our diagram (Fig. III.17), drawn from microphotographs by Hanlin, shows that conidiogenesis in *Spiniger meineckellus* is similar to that in *Poronia punctata* (Sympodulosporae, Fig. III.13); the spicules that carry the conidia are simply a little elongated here. The author considers that the outer layers of the wall of the conidiogenous cell are not found in that of the denticle (steps a and b, Fig. III.17) and that this latter is thus enteroblastic. But according to us, these outer layers are A, amorphous zone, and the interface A/B. The fundamental layer B is clearly found in the denticle, in a thin form, and in the conidium.

Conidiogenesis in the anamorph *Chromelosporium* of *Peziza ostracoderma* (Ascomycete, Pezizaceae) (studied by Hughes and Bisalputra, 1970) is absolutely the same: the authors conclude that there is no rupture of the wall of the ampulla, and that the denticles and conidia are holoblastic.

Gonatobotryum apiculatum (Fig. III.18)

The chief differences from the two preceding species are:
- the melanization of walls;
- the production of conidia in chains;
- the presence of two septa enclosing a disjunctor between conidiogenous cell and

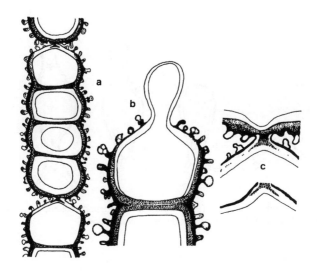

Fig. III.16. *Torula* spp. (after Ellis and Griffiths, 1975a and b).
a: general view of a conidium in a chain; **b:** detail of the fertile apical cell producing a secondary conidium; **c:** detail of relations between the basal cell of a conidium and the apical cell that gives rise to it.

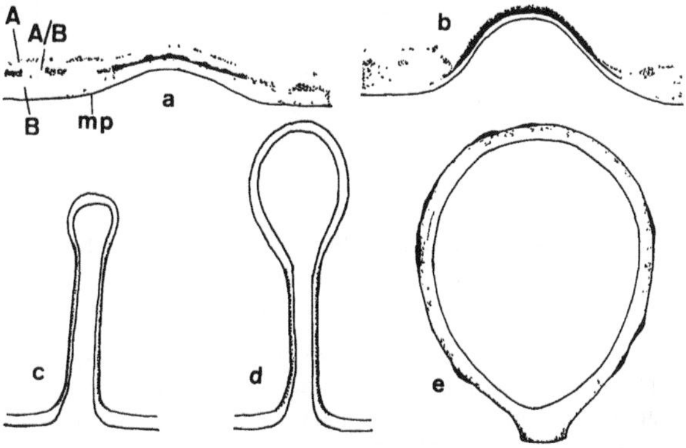

Fig. III.17. *Spiniger meineckellus*, anamorph of *Heterobasidion annosum* (Basidiomycete, Polyporaeeae) (after Hanlin, 1982).
a: thinning of the wall of the conidiogenous cell; **b:** beginning of spicule formation; **c, d:** formation of a conidium at the tip of a spicule; **e:** mature, detached conidium.
A, B: layers of the wall; **A/B:** interface of the two preceding layers; **mp:** plasmic membrane.

conidium, linked with a rhexolytic secession as in the related genus *Gonatobotrys*. *Nematogonum,* on the other hand, is schizolytic (Walker and Minter, 1980).

Conidiogenesis of the Botryoblastosporae is therefore similar to that of Sympodulosporae and Acroblastosporae.

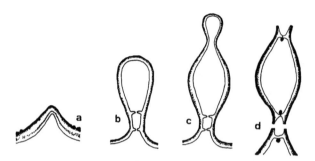

Fig. III.18. *Gonatobotryum apiculatum* (after Cole, 1973). **a:** primordium; **b:** maturation and septation; **c:** production of a secondary conidium; **d:** secession.

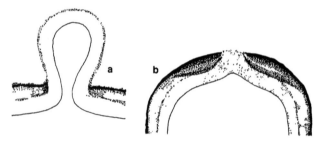

Fig. III.19. *Bipolaris maydis* (after Brotzman *et al.*, 1975).
a: conidial primordium; **b:** apex of the conidiogenous cell with pore and annular thickening.

Porosporae

Among the Porosporae that have been studied under electron microscope, we have chosen the two examples illustrated in Figs. III.19 and 20, which show that we are concerned here with a holoblastic conidiogenesis, which is not fundamentally different from that seen in the Sympodulosporae and Acroblastosporae.

It was thought that the porospores were produced in an enteroblastic manner, across the thin central channel of the pore: TEM studies show that it is not so, that the initiation of the primordium does not show a particular character, and that what is different in the Porosporae is the conidial maturation and development of the conidiogenous cell, which ends in characteristic, annular, melanized thickenings as shown by Reisinger (1970b, 1972). This author observed in *Dendryphiella vinosa* and in *Helminthosporium spiciferum* a disjunctor between conidiogenous cell and conidium, a little like that in *Gonatobotryum apiculatum* (Fig. III.18). He hypothesized that the retraction of the hilum during desiccation causes the conidial secession.

In the case of *Bipolaris maydis*, an apparent discontinuity of layers A and B2 is observed at the primordium (Fig. III.19a), a little like that in *Spiniger* at the base of the spicules. Nevertheless, Brotzman et al. (1975) considered the conidia to be holoblastic. Carroll (1971) and Carroll and Carroll (1974) concluded that there is no specific 'tretic' conidiogenesis in Porosporae, only a normal blastic process, in *Stemphylium botryosum* as in *Ulocladium atrum* (Fig. III.20).

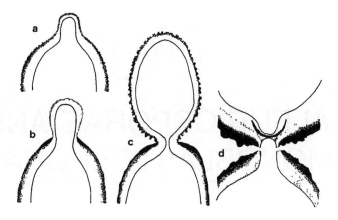

Fig. III.20. *Ulocladium atrum* (after Carroll and Carroll, 1974).
a-c: stages of conidiogenesis; **d:** secession and detail of walls at the pore.

Conclusion

We feel we have shown by the examination of ultrastructural studies that all the fungi examined here have the same type of conidiogenesis:
— holoblastic character (continuity of walls of the conidiogenous cell and the conidium);
— unique character of each conidium, each vegetative point producing only one spore.

The first criterion is in itself insufficient to differentiate the Blastosporae from Phialosporae (in which we will see that the first conidium is holoblastic). It is rather the unique character of each conidium that seems to us to be the determinant here.

Other characteristics (mode of conidiophore growth, presence or absence of chains, of pores, etc.) enable us to define a certain number of subgroups that will be examined in the following chapters, and the relations of which are established in the table below:

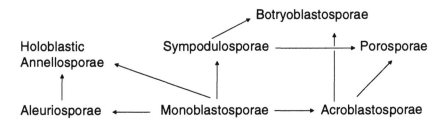

ALEURIOSPORAE AND MONOBLASTOSPORAE

The term 'aleurium' was created by Vuillemin in 1911; it designates the spores produced from the tips of fertile hyphae, or sometimes in an intercalary fashion, by bulging of the tip, appearance of a basal septum, and often thickening of the conidial wall (Fig. IV.1). For the author, the spores were released by rupture of the wall of the conidiogenous cell (rhexolytic secession).

We adopt the term 'aleuriospore' (= aleurioconidium) with the sense that Vuillemin gave to 'aleurium', but we extend it to genera in which secession is schizolytic, following the example of other authors (Tubaki, 1963; Barron, 1968). This subdivision corresponds thus to group IA of Hughes (1953): 'chlamydospores' without percurrent proliferation. The term **gangliospore**, proposed by Subramanian (1962, 1983), seems to correspond partly to aleuriospores such as those defined above, but also to chlamydospores and to arthrospores that are or are not meristematic.

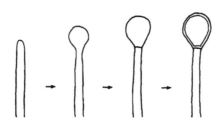

Fig. IV.1. Formation of an aleuriospore.

Sporiferous apparatus

Conidiophore, sporogenous cell

Many Aleuriosporae form conidiophores that are poorly differentiated; these are either short, lateral branches or vegetative tips of hyphae. Well-differentiated but short conidiophores are found, for example, in *Wardomyces* (Table IV.A 30). Finally, in genera such as *Staphylotrichum* (Table IV.A 40) and certain *Sporidesmium* (Table IV.B 12), they are long and melanized.

Differentiated sporogenous cells are sometimes at the tips of branches of the conidiophores (e.g.,: *Wardomyces*, Fig. III.9), becoming empty during the projection of the conidia and playing the role of disjunctor by their lateral rupture.

Conidia

In this group one finds most of the morphological types of spores (cf. Table I.C). The release of mature conidia, if it occurs by rupture of the lateral wall of the conidiogenous cell (rhexolytic secession), leaves on the hilum an annular fringe around the basal septum, very typical in *Pithomyces* (Table IV.B 24). Schizolytic secession is rarely found. Sometimes the two modes of secession coexist in the same species (e.g., *Wardomyces, Epicoccum,* Cole and Samson, 1979).

The aleuriospores present analogies with the organs of resistance known as chlamydospores, and all the intermediates may exist between the two. The chlamydospores result typically from the transformation of an intercalary hyphal cell, by bulging, accumulation of reserves at the expense of the adjacent cells, and thickening of the wall (Fig. I.16). There are also lateral and terminal chlamydospores, so that the distinction between chlamydospores and aleuriospores that are poorly differentiated is difficult: for example, the chlamydospores of *Chloridium* (of which the principal anamorph is phialidic) closely resemble certain *Humicola* (Table IV.A 24).

It therefore seems that the aleuriospore may sometimes be an accessory form of asexual multiplication, with a tendency to function as a simple organ of conservation. One frequently also finds with the aleuriospores another conidial form, a synanamorph.

• with exophialides,* chiefly in the group *Sepedonium-Mycogone* (Table IV.A 17-18);

---

*See introduction, chapter I, and Phialosporae, chapter X.

- with endophialides in certain *Chalara* with unicellular aleuriospores (earlier *Chalaropsis*, Table IV.A 37) or phragmiate aleuriospores (earlier *Thielaviopsis*, Table IV.B 4);
- finally, with annellides in the group *Echinobotryum-Wardomyces* (Table IV.A 29-30 and Table XI.A 7-4).

Ultrastructural aspects

The ultrastructural aspects of some examples of Aleuriosporae that have been studied in chapter III demonstrate that the conidia are holoblastic (conidial wall continuous with that of the conidiogenous cell).

We do not separate the holoblastic Aleuriosporae described above from rare genera with holothallic conidia known in certain Dermatophytes (Fig. I.19). In all the cases, the conidial wall is continuous with that of the conidiogenous cell, but in the latter the basal septation is formed **before** the swelling of the tip, which is the conidial primordium.

Conclusion

When a conidiogenous cell produces a unique terminal conidium, we call it aleuriospore in the Hyphales, and monoblastospore in the Conidiomales. The conditions of development and maturation of the conidia in the former cause them to have a generally characteristic appearance (wide hilum, etc.) that may not occur in the latter.

Key to identification of Aleuriosporae and Monoblastosporae

1	Hyphales	2
	Conidiomales	90

Hyphales _____

2(1)	Conidiophores and conidia poorly differentiated and variable (appearance of terminal, sometimes lateral or intercalary chlamydospores)*	3
	Conidiophores and conidia of a definite and constant form	18
3(2)	Presence of clamp connections (see also *Sporotrichum* A10)	4
	No clamp connections	6

*See also *Keratinomyces* A1 and *Trichophyton* A3.

4(3)	Two types of conidia, small sympodu-lospores and large aleuriospores	*Disporotrichum*	**A13**
	Only aleuriospores		5
5(4)	Lines of ovoid cells separated by clamps	*Ptychogaster*	**A15**
	Terminal aleuriospores, fungus more or less in the form of a yeast	*Itersonilia*	**A14**
6(3)	Conidia hyaline		7
	Conidia melanized		12
7(6)	Presence of disjunctors between conidia	*Chrysosporium*	**A9**
	No disjunctors between conidia		8
8(7)	Conidia spherical with thick walls	*Desertella*	**A22**
	Conidia not as above		9
9(8)	Mucous colonies, submerged, conidia mostly lateral	*Trichosporiella*	**A7**
	Normal aerial development of colonies		10
10(9)	Conidia globular or elongated, sometimes in short chains	*Zymonema*	**A6**
	Conidia produced on denticles		11
11(10)	Conidia lateral or terminal, smaller than 5 µm sometimes in chains	*Emmonsia*	**A11**
	Conidia larger than 5 µm, produced chiefly on ampulliform swellings	*Myceliophthora*	**A8**
12(6)	Conidia uniquely lateral, parent hyphae of sympodulosporate appearance, sporodochia in natural conditions on herbaceous plants	*Beniowska*	**A16**
	Conidia produced in other ways		13
13(12)	Appearance of large terminal chlamydospores isolated on short conidiophores		14
	Appearance of small ovoid conidia		15
14(13)	Conidia globular, smooth	*Humicola*	**A24**
	Conidia embossed or ornamented	*Thermomyces*	**A25**
15(13)	Undifferentiated, lateral conidiophores on parent hyphae		16
	Conidiophores more or less ramified		17
16(15)	Conidia ovoid	*Acremoniula*	**A26**
	Conidia cuneiform with germ pore	*Rhinocladium*	**A27**
17(15)	Ampulliform swellings	*Myceliophthora*	**A8**
	No such character	*Sporotrichum*	**A10**
18(2)	Conidial apparatus hyaline		19
	Conidial apparatus melanized		36
19(18)	Ameroconidia		20
	Conidia septate		28
20(19)	Conidia small, in ramified chains	*Geomyces*	**A12**
	Conidia larger, globular, not in chains		21
21(20)	Conidiophores short, having amero-, didymo- or phragmospores	*Fraseriella*	**B5**
	Conidiophores more developed		22

22(21)	Conidiophores erect, ending in denticles with spiny aleuriospores	23
	Conidiophores of different types	24
23(22)	Parasites of humans and animals *Histoplasma*	**A5**
	Parasites of higher fungi, phialidic synanamorph *Sepedonium*	**A17**
24(22)	Conidia globular	25
	Macroconidia cruciform and microconidia pyriform	27
25(24)	Conidiophores bulging in apical ampulla(e)	26
	Conidiophores ramified, with sterile bristles *Botryoderma*	**A20**
26(25)	Ampullae with long arms, each producing a smooth conidium *Umbelopsis*	**A21**
	Superposed ampullae with warty conidia *Glomopsis*	**A23**
27(24)	Macroconidia long, with pointed apical projections *Diademospora*	**B2**
	Macroconidia in four-leaf clover form *Riessia*	**B1**
28(19)	Conidia bicellular	29
	Conidia multicellular	31
29(28)	Conidiophore short, amero-, didymo- or phragmoconidia *Fraseriella*	**B5**
	Conidiophore larger, ramified	30
30(29)	Basal cell conical *Chlamydomyces*	**A19**
	Basal cell hemispherical *Mycogone*	**A18**
31(28)	Phragmoconidia	32
	Dictyoconidia *Miuraea*	**B6**
32(31)	Subcylindrical phragmoconidia	33
	Fusoid phragmoconidia	34
33(32)	With microconidia *Trichophyton*	**A3**
	Without microconidia, sometimes with tendrils, conidia becoming larger towards the tip *Epidermophyton*	**A4**
34(32)	Tip extending into a bristle *Spermospora*	**B3**
	No such character	35
35(34)	Macroconidia diplothecate *Keratinomyces*	**A1**
	Macroconidia haplothecate+microconidia *Microsporum*	**A2**
36(18)	Amerospores	37
	Conidia septate	51
37(36)	Conidia at tip more or less acute	38
	Conidia globular or ovoid	41
38(37)	Conidia more or less differentiated	39
	Conidia differentiated	40
39(38)	Smooth conidia with a germ slit *Mammaria*	**A28**
	Warty macroconidia, without germ slit, synanamorph *Scopulariopsis, Doratomyces* *Echinobotryum*	**A29**

40(38)	Conidiophores short, with bunches of conidia on stocky conidiogenous cells	*Wardomyces*	**A30**
	Conidiophores often bulging into an apical ampulla, with bunches of conidia on the ends of pedicels	*Asteromyces*	**A32**
41(37)	Conidia spherical		42
	Conidia obovoid or asymmetrical		44
42(41)	Conidiophores short		43
	Conidiophores long, erect, branched	*Staphylotrichum*	**A40**
43(42)	Conidiophores short, bulging	*Gilmaniella*	**A33**
	Conidiophores medium-sized, sterile bristles	*Botryotrichum*	**A41**
44(41)	Conidiophores undifferentiated		45
	Conidiophores differentiated		47
45(44)	Conidia cuneiform	*Rhinocladium*	**A27**
	Conidia obovoid		46
46(45)	Conidia sharply truncated at the base	*Acremoniula*	**A26**
	Rhexolytic separation, basal fringe on the conidium	*Pithomyces*	**B24**
47(44)	Conidiophores short		48
	Conidiophores more developed, branched		49
48(47)	Conidia small, without disjunctor	*Sporotrichum*	**A10**
	Conidia flat, with disjunctor	*Nigrospora*	**A31**
49(47)	With phialide synanamorph	*Chalara,* chap. IX earlier	
		Chalaropsis	**A37**
	Without phialide synanamorph		50
50(49)	Conidia with wide base	*Allescheriella*	**A34**
	Conidia with narrow base on pointed sporogenous cells	*Harzia*	**A35**
51(36)	Phragmoconidia		52
	Dictyo-, stauro- or helicospores		65
52(51)	Conidiophores short, undifferentiated		53
	Conidiophores more or less long, erect, differentiated		61
53(52)	Conidia with septal constrictions and apical bristle	*Arthrocladium*	**A42**
	Conidia of different types		54
54(53)	Conidia ending in a thin bristle		55
	Conidia of different types		56
55(54)	Lateral bristle at the base	*Laterispora*	**B11**
	Apical, curved bristle, hyphopodia	*Ceratophorum*	**B16**
56(54)	Conidia with rhexolytic separation		57
	Conidia with schizolytic separation		58
57(56)	Conidia 0- to multiseptate, sometimes dictyate, very variable according to species	*Pithomyces*	**B24**
	Conidia fusoid, diplothecate	*Murogenella*	**B14**

58(56)	Conidia diplothecate, fusoid	*Murogenella*	**B14**
	Conidia of different types		59
59(58)	Conidia elongated, rarely cylindrical, most often attenuated towards the tip	*Sporidesmium*	**B12**
	Conidia shorter, pyriform or obovoid		60
60(59)	Very short or no conidiophores on hyphae	*Trichocladium*	**B13**
	Conidiophores semimacronemous, grouped in sporodochia, conidia often diplothecate or with opaque, more or less significant bands	*Bactrodesmium*	**B23, C6**
61(53)	With phialide synanamorph *Chalara*, chap. X, aleuriospores breaking up into individual pieces	earlier *Thielaviopsis*	**B4**
	No such character		62
62(61)	Conidia cylindrical, diplothecate	*Henicospora*	**B15**
	Conidia different		63
63(62)	Conidia tapered at the tip in a hook	*Uncispora*	**B25**
	Conidia fusoid, ovoid		64
64(63)	Conidia long and often pointed	*Sporidesmium*	**B12**
	Conidia of variable septation (0- to multiseptate and even dictyate) and coloration (often multicoloured), conidiophore with recurrent growth, sometimes branched	See *Endophragmiella* chap. V B10)	
65(51)	Dictyospores		66
	Stauro- or helicospores		82
66(65)	Conidia hyaline	*Miuraea*	**B6**
	Conidia at least partly melanized		67
67(66)	Conidia formed from one central melanized cell and from hyaline satellite cells		68
	No such character		70
68(67)	Satellite conidia in a helicoidal arrangement	*Acrospeira*	**A38**
	Satellite conidia in a symmetrical arrangement		69
69(68)	Two lateral satellite cells, conidiogenous cells verticillate on long conidiophores	*Physalidiella*	**A43**
	Basal, apical and lateral satellite cells	*Stephanoma*	**A36**
70(67)	Conidia discoloured with hyaline and melanized cells, sometimes opaque bands		71
	No such character		72
71(70)	Conidia more or less lenticular or flat	*Oncopodium*	**B7, C9**
	Conidia triangular-shaped	*Oncopodiella*	**B8**
72(70)	Conidia small or medium-sized		73
	Conidia large		78

73(72)	Conidia globular or flat		74
	Conidia ovoid		76
74(73)	Conidia flat	*Tetracoccosporium*	**B29**
	Conidia more or less isodiametrical		75
75(74)	Conidia spiny and more or less spherical (in sporodochia in a natural substrate)	*Epicoccum*	**B31, C10**
	Conidia smooth and more or less irregular-shaped, phialidic synanamorph	*Diheterospora*	**A39**
76(73)	Longitudinal and oblique septa in the apical and sometimes basal cell	*Bactrodesmium*	**B23**
	Longitudinal septa in the body of the conidium		77
77(76)	Rhexolytic secession	*Pithomyces*	**B24**
	Schizolytic secession	*Monodictys*	**B30**
78(72)	Conidia homogenous, uniformly coloured		79
	Conidia discoloured with heterogeneous elements		81
79(78)	Conidia fan-shaped	*Mycoenterolobium*	**B26**
	Conidia elongated		80
80(79)	Conidia muriform	*Berkleasmium*	**B18**
	Conidia that seem to be formed from a group of joined phragmospores	*Dictyosporium*	**B19**
81(78)	Conidia ovoid with phialides, presence of sterile bristles	*Septosporium*	**B17**
	Conidia growing with one or several spherical, secondary conidia	*Xenosporium*	**B33**
82(65)	Conidia unicellular (see *Riessia, Diademospora*)		27
	Conidia multicellular		83
83(82)	Helicospores	*Hobsonia*	**B32**
	Staurospores		84
84(83)	Conidiophores short		85
	Conidiophores long		88
85(84)	Conidia that seem to be formed from four joined phragmospores ending in diverging arms	*Tetraploa*	**B28**
	Conidia of different types		86
86(85)	Branches at the base of the conidia		87
	Conidia of an irregular star shape	*Tripospermum*	**B22**
87(86)	Rounded base of the conidium	*Ceratosporium*	**B20**
	Pointed base of the conidium	*Digitodesmium*	**B10**
88(84)	Conidia with drooping arms	*Cryptocoryneum*	**B9**
	Conidia with arms open towards the top		89
89(88)	A single di- or trichotomous branch at the base of the conidium	*Ceratosporella*	**B21**
	Several dichotomous branches	*Anavirga*	**B27**

Conidiomales

90(1)	Conidiophores joined in loose bundles or in a coremium		91
	Sporodochia, acervuli, cupules or pycnidia		95

Bundles or coremia

91(90)	Coremium		92
	Bundles of conidiophores		93
92(91)	Coremium with conidia on the outside in the shape of a four-leaf clover	*Riessia*	**B1**
	Coremium with phragmoconidia in an apical cupule	*Morrisographium*	**C1**
93(91)	Hyaline conidia		94
	Melanized helicospores	*Troposporella*	**C2**
94(93)	Phragmoconidia	*Bactridium*	**C3**
	Botryoconidia	*Cheirospora*	**C4**
95(90)	Sporodochia		96
	Other conidiomata		104

Sporodochia

96(95)	Phragmoconidia		97
	Dictyo- or stauroconidia		98
97(96)	Conidia small, light-coloured	*Pollaccia*	**C5**
	Conidia large, dark	*Bactrodesmium*	**C6**
98(96)	Dictyoconidia		100
	Stauroconidia		99
99(98)	Thin, entirely fertile sporodochia, conidia in multiple, descending branches	*Cryptocoryneum*	**C12**
	Massive sporodochia with dispersed fertile zones, conidia in multiple ascending or radiating branches	*Atichia*	**C13**
100(98)	Elongated dictyoconidia		101
	Globular or flat dictyoconidia		102
101(100)	Fusoid form, lighter, elongated tip	*Dictyodesmium*	**C7**
	Generally cylindrical form	*Berkleasmium*	**C8**
102(100)	Conidia small, spherical, uniformly coloured	*Epicoccum*	**B31, C10**
	Conidia of non-homogeneous constitution and coloration		103
103(102)	Conidia lenticular, wide basal ampulla, hyaline cells at the extremities and central, dark band present or absent depending on the species	*Oncopodium*	**C9**
	Conidia bristled	*Petrakia*	**C11**
104(95)	Acervuli or cupules		105
	Pycnidia or pseudopycnidia		127
105(104)	Acervuli		106
	Cupules		120

Acervuli

106(105) Ameroconidia		107
Conidia septate		111
107(106) Conidia hyaline		108
Conidia melanized		110
108(107) Conidia bristled	*Chaetospermum*	**C14**
Conidia muticate		109
109(108) Simple conidiogenous cell	*Leptodermella*	**C15**
Superposed conidiogenous cells	*Catenophora*	**C16**
110(107) Conidia ovoid	*Gaubaea*	**C17**
Conidia turbinate with a light line under the upper crown	*Scyphospora**	**C18**
111(106) Didymoconidia		112
Phragmo-, dictyo- or stauroconidia		113
112(111) Conidia retracted at the septum, hyaline on a typical acervulus	*Marssonnina*	(V D1)
Conidia not retracted, light brown, on bursting sporodochia	*Oncospora*	**C19**
113(111) Phragmoconidia with hyaline end cells	*Scolicosporium*	**C20**
Dictyo- or stauroconidia		114
114(113) Dictyoconidia		115
Stauroconidia		117
115(114) Hyaline dictyoconidia	*Thyrsidina*	**C21**
Melanized dictyoconidia with mucilaginous sheath		116
116(115) Dictyoconidia s.s.	*Myxocyclus*	**C22**
Conidia botryoid	*Endobotrya*	**C23**
117(114) Melanized stauroconidia	*Asterosporium*	**C24**
Hyaline stauroconidia		118
118(117) Conidia in the form of a many-pointed star	*Psammina*	**C25**
Other forms		119
119(118) Insect-like form	*Entomosporium*	**C26**
Form of cat's paw	*Vestigium*	(VI H9)

Cupules

120(105) Didymoconidia		121
Phragmo-, dictyo- and stauroconidia		122
121(120) Cupules with thick walls, elongated conidia	*Pseudocenangium*	(V D25)
Cupules with thin walls, ovoid conidia	*Cystotricha*	**C27**
122(120) Phragmoconidia		123
Dictyo- or stauroconidia		124
123(122) Small cupules with marginal bristles	*Protostegia*	**C28**
Large cupules with thick walls	*Fujimyces*	**C29**

---

*Culture studies showed (Samuels et al., 1981) that, in reality, the genus *Scyphospora* belongs to the group of basauxic Deuteromycetes (cf. chap. XII).

124(122)	Conidia botryoid	*Endobotryella*	**C30**
	Stauroconidia		125
125(124)	Hyaline stauroconidia	*Prosthemium*	**C31**
	Melanized stauroconidia		126
126(125)	Star-shaped conidia	*Asterosporium*	**C24**
	Conidia in the form of fingers arising		
	from a common base	*Sirothecium*	**C32**
127(104)	Pycnidia		128
	Pseudopycnidia		160

Pycnidia

128(127)	Conidioma superficial		129
	Immersed or semi-immersed conidioma		132
129(128)	Superficial mycelium with bristles	*Chaetodiplodina*	**D1**
	No such character		130
130(129)	Conidia hyaline with apical appendages,		
	pycnidia not ostiolate	*Japonia*	**D2**
	Conidia melanized		131
131(130)	Phragmoconidia, hyaline apical cells		
	with appendages	*Labridella*	**D3**
	Hypophyllous, non-ostiolate pycnidia,		
	with a star-shaped or irregular opening;		
	conidia with a hyaline median band	*Capnodiastrum*	**D4**
132(128)	Pale yellow, rostrate conidioma, hyaline		
	amerospores sprouting sometimes in a		
	cross shape in the rostrum	*Hyalopycnis*	**D5**
	Melanized conidioma		133
133(132)	Conidia with appendages		134
	Conidia without appendages		146
134(133)	Appendage(s) in apical position		135
	Amphigenous appendages		144
135(134)	A single apical appendage		136
	Several apical appendages		140
136(135)	Cellular appendage		137
	Mucilaginous appendage		138
137(136)	Conidia hyaline	*Kellermania*	**D6**
	Conidia brown	*Scolecosporiella*	**D7**
138(136)	Conidia unicellular		139
	Conidia multicellular	*Neottiosporina*	**D9**
139(138)	Mucilaginous sheath, tapering		
	appendage	*Phyllosticta*	**D15**
	Appendage funnel-shaped	*Tiarosporella*	**D8**
140(135)	Conidia hyaline		141
	Conidia brown		143
141(140)	Conidia unicellular		142
	Conidia multicellular	*Alpakesa*	**D10**
142(141)	Appendages mucilaginous	*Tiarosporella*	**D8**
	Appendages: fine bristles	*Giulia*	**D11**

143(140)	Appendages non-ramified, conidiophores elevated	*Hyalotiella*	**D12**
	Appendages ramified, compact conidiogenous cells	*Hyalotiopsis*	**D13**
144(134)	Appendages cellular	*Scopaphoma*	**D16**
	Appendages mucilaginous		145
145(144)	Conidia unicellular	*Pseudobasidiospora*	**D22**
	Conidia bicellular	*Tiarospora*	**D14**
146(133)	Conidia hyaline		147
	Conidia brown		155
147(146)	Conidia variable, pycnidia aristate	*Aristastoma*	**D23**
	Pycnidia without bristles		148
148(147)	Pycnidia with a clypeus (or a clypeiform tip)		149
	Pycnidia without clypeus		150
149(148)	Clypeus, septate macroconidia and amerosporous microconidia	*Clypeostagonospora*	**D28**
	Ostiole topped with a 'chimney'	*Cytostagonospora*	**D20**
150(148)	Conidia unicellular	*Phyllosticta*	**D15**
	Conidia septate		151
151(150)	Conidia pyriform, bicellular	*Apiocarpella*	**D17**
	Conidia phragmiate		152
152(151)	Conidia filiform	*Jahniella*	**D21**
	Conidia wider		153
153(152)	Conidioma with thin wall	*Dearnessia*	**D18**
	Conidioma with thick wall		154
154(153)	Conidioma unilocular with dark outer wall and hyaline inner wall	*Stagonospora*	(V C4)
	Conidiomata uni- or multilocular with uniformly dark wall	*Piptarthron*	**D19**
155(146)	Conidia muriform, distoseptate	*Camarosporellum*	**D24**
	Conidia euseptate		156
156(155)	Conidia non-muriform		157
	Conidia muriform		159
157(156)	Conidia bicellular		158
	Conidia spiny, multicellular	*Sclerostagonospora*	**D27**
158(157)	Equal cells	*Diplodia*	**D25**
	Bulging apical cell	*Kamatella*	**D26**
159(156)	Conidia ellipsoid	*Stigmella*	**D29**
	Conidia subspherical	*Nummospora*	**D30**

Pseudopycnidia

160(127)	Conidiogenous cells with clamp connections		161
	No such character		163
161(160)	Pseudopycnidium cupuliform at tip of a massive stromatic foot	*Cenangiomyces*	**E6**
	Pseudopycnidium underlying, unilocular		162

(Continued on p. 92)

Table IV.A. Hyaline Hyphales, Aleuriosporae

Fig. Genus	No. spp.	Geographical distribution	Mode of life, substrates and impact	Teleomorphs	Additional references	Mol. biol.
1 *Keratinomyces*	1	Cosmopolitan	Telluric, dermatophyte	*Arthroderma*		
2 *Microsporum*	14	Cosmopolitan	Telluric, keratinophile, humans, animals	*Arthroderma*		X
3 *Trichophyton*	20	Cosmopolitan	Telluric, keratinophile, humans, animals	*Arthroderma*		X
4 *Epidermophyton*	2	Cosmopolitan	Dermatophyte on hairless skin (human)			X
5 *Histoplasma*	2	Ubiquitous	Telluric, underlying mycosis (humans and animals)	*Ajellomyces*		X
6 *Zymonema*	1	N. and S. Amer., Africa	Blastomycosis in humans	*Ajellomyces*	Stud. Mycol. 20 (1980)	
7 *Trichosporiella*	4	Ubiquitous	Telluric	*Laetinaevia*	Stud. Mycol. 20 (1980)	
8 *Myceliophthora*	9	Cosmopolitan	Telluric, keratinolytic (humans and animals), plant debris	*Arthroderma* *Corynascus* *Ctenomyces*	Stud. Mycol. 20 (1980)	X
9 *Chrysosporium*	40	Cosmopolitan	Telluric, keratinophile, cellulolytic and/or pathogenic on humans and some animals, coprophile, glucophile	*Aphanoascus* *Arthroderma* *Gymnoascus* and other Onygenales	Stud. Mycol. 20 (1980)	X
10 *Sporotrichum*	3	Cosmopolitan	Telluric, lignicolous, underlying mycosis (humans), toxinogenic	*Laetiporus* (B) *Phanerochaete* (B), *Pycnoporellus* (B)	Stud. Mycol. 24 (1984)	X
11 *Emmonsia*	1	Cosmopolitan	Telluric, underlying mycosis (humans, rodents)		Stud. Mycol. 20 (1980)	
12 *Geomyces*	4	Ubiquitous, mostly temperate North, Australia	Telluric	*Pseudogymnoascus*	Stud. Mycol. 20 (1980) Mycol. 80:82 (1988)	
13 *Disporotrichum*	1	Eurasia	Air and various material		Stud. Mycol. 24 (1984)	
14 *Itersonila*	3	Temperate zone, Central Amer.	Atmospheric pollutant, parasite on root of *Pastinaca* and on various flowers			
15 *Ptychogaster*	10	Ubiquitous	Lignolytic	*Tyromyces* (B), *Oligoporus* (B)		
16 *Beniowskia*	2	Mostly tropical zone	Leaf parasite on *Poaceae*			
17 *Sepedonium*	10	Temperate zone	Telluric, mycoparasite	*Apiocrea*, *Hypomyces*, *Thielavia*		
18 *Mycogone*	11	Cosmopolitan	Mycoparasite	*Hypomyces*		
19 *Chlamydomyces*	2	Cosmopolitan	On plants			
20 *Botryoderma*	3	Southern hemisphere	Telluric			
21 *Umbelopsis**	1	N. Amer.	Telluric			
22 *Desertella*	1	Egypt	Telluric			
23 *Glomopsis*	2	N. Amer.	Leaf parasite (*Cornus, Lonicera*)	*Herpobasidium* (B)		

*The type species of the genus is in fact an anamorph of Zygomycete; each 'conidium' is a unisporate sporangiole (Mycot. 51: 15, 1994).

Hyaline Hyphales, Aleuriosporae

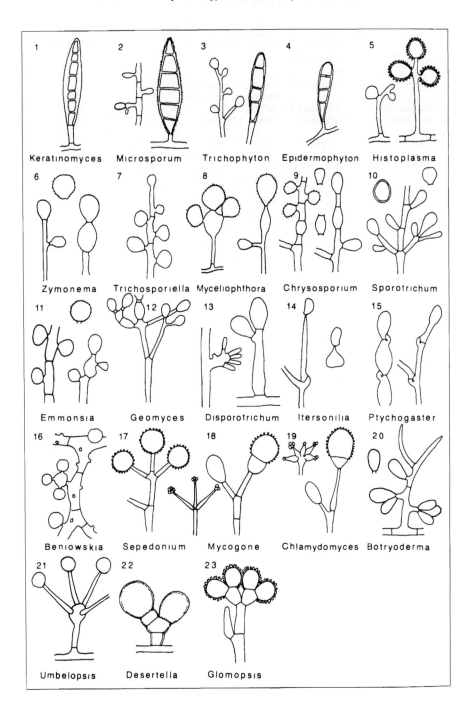

Table IV.A. (Contd.) Melanized Hyphales, Aleuriosporae

Fig.	Genus	No. spp.	Geographical distribution	Mode of life, substrates and impact	Teleomorphs	Additional references	Mol. biol.
24	*Humicola*	± 30	Cosmopolitan	Telluric			X
25	*Thermomyces*	4	Cosmopolitan	Telluric			X
26	*Acremoniula*	4	Cosmopolitan	Telluric, mycoparasite, lignicolous, detriticolous			
27	*Rhinocladium*	5	Cosmopolitan	Telluric			
28	*Mammaria*	1	Temperate North	Telluric, lignicolous			
29	*Echinobotryum*	1	Temperate North	Telluric, lignicolous, detriticolous, fimicolous			
30	*Wardomyces*	8	Cosmopolitan	Telluric, detriticolous, contaminant of food	*Microascus*		
31	*Nigrospora*	5	Cosmopolitan mostly tropical zone	Telluric, plant pathogen (*Zea, Oryza....*)	*Khuskia*		
32	*Asteromyces*	1	Temperate Europe	Sand dunes			
33	*Gilmaniella*	5	Cosmopolitan	Telluric, fimicolous, 1 sp. halophilic		Myc. Res. 92:502 (1989)	
34	*Allescheriella*	2	Cosmopolitan	Lignicolous	*Botryobasidium* (B)		
35	*Harzia*	5	Cosmopolitan	Telluric, lignicolous, seminicolous, detriticolous	*Melanospora*		
36	*Stephanoma*	3	N. Amer., Eurasia	Telluric, radicicolous, mycoparastie	*Hypomyces*		
37	*Chalaropsis*	6	Cosmopolitan	Plant parasite	*Ceratocystis*		
38	*Acrospeira*	1	Temperate North	Parasite on chestnut			
39	*Diheterospora*	12	Cosmopolitan	Telluric, fimicolous, radicicolous, parasite of rotifers, of eggs, of gasteropods, and of nematode cysts		CJB, 63: 211 (1985)	
40	*Staphylotrichum*	1	Cosmopolitan	Telluric			
41	*Botryotrichum*	4	Cosmopolitan	Telluric, fimicolous, nematophagous	*Chaetomium*	Mycot. 48: 271 (1993)	
42	*Arthrocladium*	1	S. Africa	Telluric			
43	*Physalidiella*	2	Ubiquitous	Telluric and on culm of *Triticum*			

Melanized Hyphales, Aleuriosporae

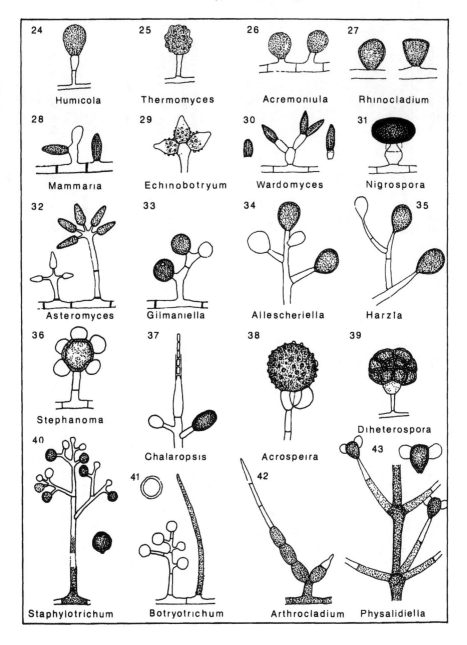

24 Humicola
25 Thermomyces
26 Acremoniula
27 Rhinocladium
28 Mammaria
29 Echinobotryum
30 Wardomyces
31 Nigrospora
32 Asteromyces
33 Gilmaniella
34 Allescheriella
35 Harzia
36 Stephanoma
37 Chalaropsis
38 Acrospeira
39 Diheterospora
40 Staphylotrichum
41
42 Botryotrichum
Arthrocladium
43 Physalidiella

Table IV.B. Hyphales, Aleuriosporae

Fig.	Genus	No. spp.	Geographical distribution	Mode of life, substrates and impact	Teleomorphs	Additional references	Mol. biol.
1	Riessia	4	N. and S. Amer., Eurasia	Lignicolous	Basidiomycete		
2	Diademospora	1	Sweden	Forest soil			
3	Spermospora	6	Ubiquitous	On herbaceous plants			
4	Thielaviopsis	2	Cosmopolitan	Root pathogen or parasite on various organs of many plants	Ceratocystis		
5	Fraseriella	1	Australia, Italy	On liquorice stick	Xeromyces		
6	Miuraea	3	Cosmopolitan	Minor pathogen on fruits of Rosaceae	Mycosphaerella		
7	Oncopodium	4	Cosmopolitan	Saprobe			
8	Oncopodiella	1	Europe	Bark, dead wood			
9	Cryptocoryneum	4	Europe	Bark, dead wood			
10	Digitodesmium	1	Great Britain	Dead wood			
11	Laterispora	1	USA	Sclerotia of Sclerotinia			
12	Sporidesmium	±70	Cosmopolitan	Saprophyte on various plants	Eupelte		
13	Trichocladium	26	Cosmopolitan	Telluric, aquatic, detriticolous	Gaeumannomyces, Halosphaeria		
14	Murogenella	2	N. Amer., India	Telluric, corticolous			
15	Henicospora	4	Cosmopolitan	Saprobe			
16	Ceratophorum	2	Europe, USA	Hypophyllous on leaves of Quercus			
17	Septosporium	5	Cosmopolitan	Saprobe			
18	Berkleasmium	14	Cosmopolitan	Lignicolous		Mycot. 34: 474 (1989)	
19	Dictyosporium	24	Cosmopolitan	Telluric, lignicolous and on blades of grass, leaves		Myc. Res. 95: 1145 (1991)	
20	Ceratosporium	4	Cosmopolitan	Dead wood, bark, branches	Iodosphaeria		
21	Ceratosporella	8	Cosmopolitan	Dead wood, bark, branches. leaves		Myc. Res. 95: 158 (1991)	
22	Trispermum	9	Cosmopolitan	Dead leaves and dead stems	Trichomerium		
23	Bactrodesmium	20	Cosmopolitan	Wood and bark of deciduous and coniferous trees	Stuartella		
24	Pithomyces	22	Cosmopolitan	Telluric, detriticolous, atmospheric pollutant	Leptosphaerulina	Stud. Mycol. 28 (1986)	
25	Uncispora	1	USA	Saprobe			
26	Mycoenterolobium	1	Ubiquitous, mostly tropical	Telluric, detriticolous			
27	Anavirga	3	Eurasia	Saprobe	Vibrissea		
28	Tetraploa	6	Cosmopolitan	Aquatic and detriticolous	Massarina		
29	Tetracoccosporium	4	Europe, India	Telluric, fimicolous			
30	Monodictys	28	Cosmopolitan	Soil, air, sea, various plant debris	Ohleria, Tubeufia	Stud. Mycol. 28 (1986)	
31	Epicoccum	1	Cosmopolitan	Very frequent on various materials, toxinogenic			
32	Hobsonia	3	Cosmopolitan	Lichenicolous 1 sp. on Phanerogams			
33	Xenosporium	12	Ubiquitous, mostly tropical	Various plant debris	Herpotrichia, Thaxteriellopsis	Mycol. 82: 742 (1990)	

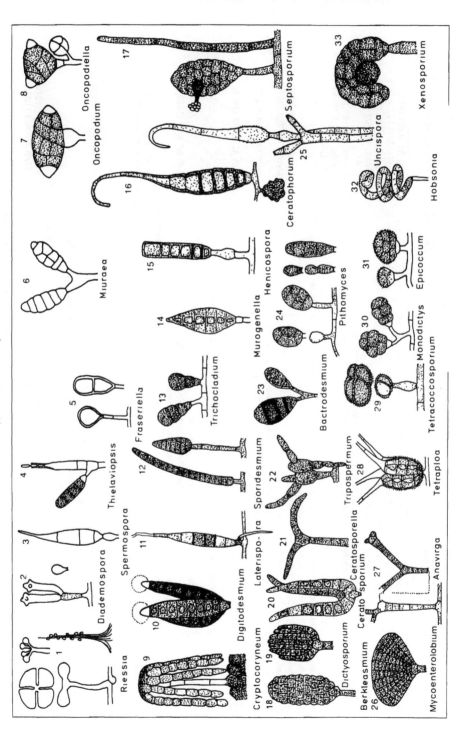

Table IV.C. Stromatic Monoblastosporae

Fig.	Genus	No. spp.	Geographical distribution	Mode of life, substrates and impact	Teleomorphs	Additional references	Mol. biol.
1	*Morrisographium*	7	Temperate North, tropics	Lignicolous, corticolous, on dead (and sometimes living) branches and twigs of various plants		CJB 63: 423 (1985); S.A.N. 137: 573 (1990)	
2	*Troposporella*	1	Temperate North	Corticolous		Myc. Pap. 132 (1973)	
3	*Bactridium*	18	Cosmopolitan	Lignicolous, detriticolous		Mycot. 14: 227 (1982)	
4	*Cheirospora*	1	Ubiquitous	On twigs (*Quercus, Fagus, Cornus, Hedera*)			
5	*Pollaccia*	7	Temperate North	Cause of spot on leaves and shoots of Salicaceae (scab)	*Venturia*	Crypto. Myc. 6: 101 (1985); Myc. Res. 99: 983 (1995)	
6	*Bactrodesmium*	20	Cosmopolitan	Wood and bark of deciduous and coniferous trees, paper	*Stuartella*		
7	*Dictyodesmium*	2	Temperate North	Foliicolous (parasite or saprophyte)			
8	*Berkleasmium*	14	Cosmopolitan	Lignicolous, detriticolous		Mycot. 34: 474 (1989)	
9	*Oncopodium*	4	Cosmopolitan	Saprobe, detriticolous		Mycot. 70: 793 (1978)	
10	*Epicoccum*	1	Cosmopolitan	Very frequent on various materials, toxinogenic			
11	*Petrakia*	2	Europe	Maculicolous on leaves and dead branches of *Acer*		ABN 171: 221 (1968)	
12	*Cryptocoryneum*	4	Europe	Dead wood, bark			
13	*Atichia*	6	Europe, tropical zone	Leaf parasite	*Seuratia*	CJB 53: 2483 (1975)	
14	*Chaetospermum*	4	Cosmopolitan	Various plant organs (*Alnus, Cupressus, Dahlia, Epilobium, Salix...*)			
15	*Leptodermella*	1	Europe	Dead stems of *Matricaria*			
16	*Catenophora*	3	USA	Leaves and stems (*Garrya, Prunus, Yucca*)			
17	*Gaubaea*	2	Eurasia	Dead stems (*Calligonum, Atraphaxis*)			
18	*Scyphospora*	5	Eurasia	Stems (Bamboo)	*Apiospora*		
19	*Oncospora*	1	Ubiquitous	Corticolous, foliicolous	*Dermea*	CJB 66: 898 (1988)	
20	*Scolicosporium*	2	Europe	Twigs and bark (*Fagus, Populus*)	*Asteromassaria*	BSMF 106: 107 (1990)	
21	*Thyrsidina*	1	Europe	Dead branches of *Acer*		Mycot. 42: 467 (1991)	
22	*Myxocyclus*	1	N. Amer., Europe	Stems and bark of *Betula*	*Splanchnonema*		
23	*Endobotrya*	1	N. Amer.	Branches of *Fagus*			
24	*Asterosporium*	4	Temperate zone	Twigs of *Acer, Fagus, Betula*			
25	*Psammina*	2	Europe	Leaves (*Juncus, Phragmites...*)			
26	*Entomosporium*	1	Cosmopolitan	Maculicolous parasite on leaves and fruits of woody Rosaceae	*Diplocarpon*	Mik. Fit. 26: 27 (1992)	
27	*Cystotricha*	1	Europe	Lignicolous	*Xylogramma*		
28	*Protostegia*	1	S. Africa	On living leaves of *Euclea* spp.			
29	*Fujimyces*	1	Europe	Dead leaves and cones of *Pinus*		Mycol. 71: 918 (1979)	
30	*Endobotryella*	1	Europe	Bark of *Pinus*		TBMS 75: 434 (1980)	
31	*Prosthemium*	5	Ubiquitous	Twigs of *Alnus, Betula, Tilia*, leaves of *Kentia*, rotting wood	*Pleomassaria*		
32	*Sirothecium*	2	Europe	Old wood (*Populus, Pinus*)			

Stromatic Monoblastosporae

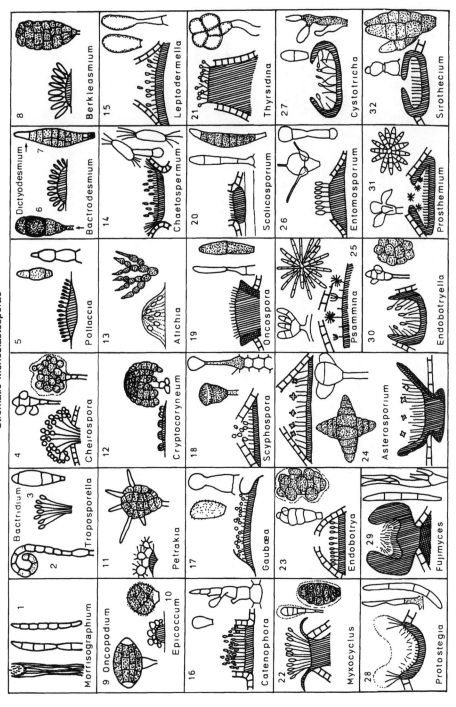

Table IV.D. Pycnidial Monoblastosporae

Fig	Genus	No. spp.	Geographical distribution	Mode of life. substrates and impact	Teleomorphs	Additional references	Mol. biol
1	Chaetodiplodina	2	S. Amer.	Folicolous	Yoshinagaia		
2	Japonia	1	Japan	On leaves of Quercus	Griphosphaenoma		
3	Labridella	1	N Amer	On twigs of Symphoricarpus	Rhytidengleruia		
4	Capnodiastrum	12	S Amer . India	On leaves of various species	Heterogastridium (B)		
5	Hyalopycnis	1	N Amer.. Europe	Parasite on carpophores of Agaricales	Planistromella	Mycot 55. 255 (1995)	
6	Kellermania	3	USA	On leaves of Agaraceae		CJB 46: 189 (1968)	X
7	Scolecosporella	6	Ubiquitous	On living or dead leaves of various plants	Leptosphaena. Phaeosphaeria		
8	Tiarosporella	8	Ubiquitous	On needles (Abies, Picea, Pseudotsuga), leaves and twigs of Poaceae	Darkera	Myc Res 99 832 (1995)	
9	Neottriosponna	10	Ubiquitous	Parasite of leaves and twigs (on Cyperaceae, Poaceae)			
10	Alpakesa	4	USA	On leaves of Yucca and Nolina	Planstromella	CJB 50 877 (1972)	
11	Giula	1	Italy	Dead stems of Lepidium graminifolium			
12	Hyalotiella	2	Ubiquitous, mostly tropical	On leaves (Acacia, Eucalyptus, Gleditschia)			
13	Hyalotiopsis*	1	India	Bark and dead leaves of Tamarindus	Ellurema		
14	Tiarospora	2	Ubiquitous	On culm of Ammophila and leaves of Deschampsia	Phaeosphaera		
15	Phyllosticta	59	Ubiquitous	Parasite often maculicolous on living leaves of many leafy or woody plants (deciduous or coniferous), also endophyte	Guignardia	Syd 43 148 (1991)	
16	Scopaphoma**	1	N Amer.	On carpophores of Coriolus and Polyporus			
17	Apiocarpella	8	Cosmopolitan	Leaf parasite		Fitol 29 39 (1985)	
18	Dearnessia	1	Canada	Leaf necrosis of Apocynum			
19	Piptarthron	6	N Amer., Europe	Leaf parasite of Agavaceae	Planistroma	Mycot 55 255 (1995)	
20	Cytostagonospora	2	Australia	On dead leaves (Arbutus. Photinia)		TBMS 87 99 (1986)	
21	Jahniella	3	Eurasia	On dead stems (Campanula, Scrophularia) and dead leaves (Hordeum)			
22	Pseudobasidiospora	1	N Amer., Australia	Rotting leaves of angiosperms			
23	Aristastoma	5	Ubiquitous	Leaf necrosis (Dolichos. Setaria, Phaseolus, Vigna)			
24	Camarosporellum	3	Europe. USA	Folicolous. maculicolous (Cercocarpus, Magnolia, Prunus)		Myc Pap 145 (1981)	
25	Diplodia	± 30	Cosmopolitan	Parasite for saprobic on various organs of many plants	Botryosphaeria, Cucurbitaria, Massarina. Otthia, Rhytidhysteron		
26	Kamatella	1	S. Africa, India	Leaf parasite on Eugenia			
27	Sclerostagonospora	1	Europe	Dead stems of Heracleum			
28	Clypeostagonospora	1	USA	Black leaf necrosis of Muhlenbergia			
29	Stigmella	26	Ubiquitous	Essentially a leaf parasite (Quercus spp...)			
30	Nummospora	1	Switzerland	On dead leaves of Carex			

*Some percurrent growths exist in this species.
**See also Bartalinia V C3.

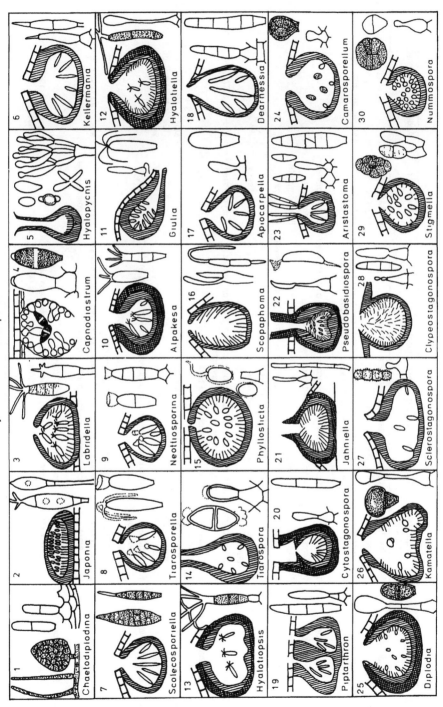

Pycnidial Monoblastosporae

Table IV.E. Pseudopycnidial Monoblastosporae

Fig.	Genus	No. spp.	Geographical distribution	Mode of life, substrates and impact	Teleomorphs	Additional references	Mol. biol.
1	*Fibulocoela*	1	Cuba, India	On rotting leaves of *Bambusa, Bucida*	Basidiomycete		
2	*Ellula*	1	Brazil	On culm of *Gadua*	Basidiomycete		
3	*Harknessia*	29	Cosmopolitan	Various organs of *Eucalyptus*, dead leaves and twigs (*Araucaria, Quercus, Rhus...*), needles (*Chamaecyparis, Juniperus*), bark (*Thuja*)	*Mebarria, Wuestneia*	Mycol. 85: 108 (1993)	
4	*Digitosporium*	1	Europe	Cause of canker on branches and trunk of pine (*Pinus*)	*Crumenulopsis*		
5	*Rabenhorstia*	10	Europe	1 sp. frequent on dead branches of *Tilia*	*Hercospora*		
6	*Cenangiomyces*	1	England	Saprophyte on needles of *Pinus*	Basidiomycete		
7	*Hysterodiscula*	2	Europe	1 sp. parasite on stems of *Empetrum* (= tar spot disease)	*Duplicaria*		
8	*Cymbothyrium*	1	France	On stem of *Smilax aspera*			
9	*Polymorphum*	1	N. Amer., Eurasia	On trunk (*Fagus, Quercus*)	*Ascodichaena*		
10	*Fusicoccum*	± 50	Cosmopolitan	Parasite (canker, drying out) or saprophyte on various organs of many plants, endophyte	*Botryosphaeria*	RFF 45: 37 (1983)	X
11	*Endomelanconium*	2	Europe, S. Amer.	On dead branches and bark of *Abies*	*Austrocenangium*		
12	*Aplosporella*	44	Cosmopolitan mostly tropical zone	On stems and dead twigs of many plants		Mycot. 45: 389 (1992); N. Hedw. 60: 79 (1995)	
13	*Lasiodiplodia*	1	Ubiquitous mostly tropical zone	Weak parasite on many plants, cause of keratitis in humans	*Physalospora*		X
14	*Placothyrium*	1	Germany	On dead stems of fern (*Athyrium*)			
15	*Dothistroma*	1	Cosmopolitan	Pathogen on needles of *Pinus* spp. (= red band disease)	*Eruptio*	Myc. Pap. 153 (1984)	X
16	*Kabatia*	9	N. Amer., Eurasia	Leaf necrosis chiefly on *Lonicera* spp.	*Discosphaerina*		
17	*Pleurothyrium*	1	Europe	On dead stems of fern (*Athyrium*)			
18	*Discosia*	31	Cosmopolitan	Dead and moribund leaves of various trees and shrubs (*Eucalyptus, Fagus, Populus, Quercus*), scales of cones (*Abies*)		Mycot. 49: 199 (1993)	
19	*Septoriella*	6	Ubiquitous	Stems, leaves and culm (*Bambusa, Juncus, Phragmites*)	*Phaeosphaeria*		
20	*Dichomera*	40	Ubiquitous	Parasite or saprobe on various organs of many plants	*Cucurbitaria*		

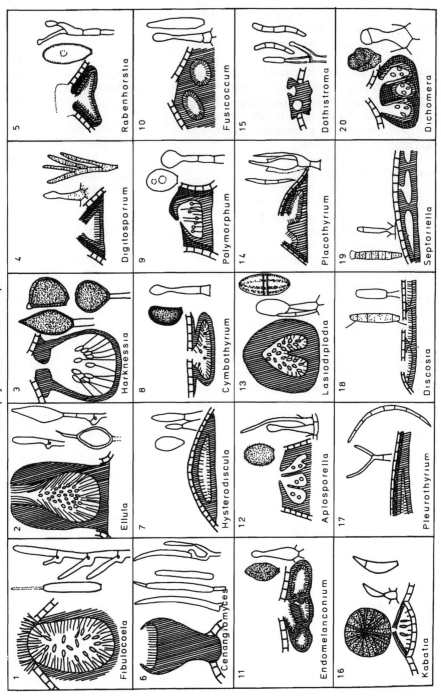

Pseudophycnidial Monoblastosporae

1 Fibulocoela
2
3 Harknessia
4 Digitosporium
5 Rabenhorstia
6 Cenangiomyces
7 Ellula
8 Cymbothyrium
9 Polymorphum
10 Fusicoccum
11 Endomelanconium
12 Hysterodiscula
13 Lasiodiplodia
14 Placothyrium
15 Dothistroma
16 Kabatia
17 Pleurothyrium
18 Discosia
19 Septoriella
20 Dichomera

162(161)	Conidioma immersed, non-ostiolate	*Fibulocoela*	**E1**
	Conidioma largely erumpent, ostiolate	*Ellula*	**E2**
163(160)	Conidioma hemispherical		
	(see also *Hysterodiscula*)		164
	No such character		165
164(163)	Conidia uni- to tricellular, non-ramified	*Kabatia*	**E16**
	Conidia multicellular, ramified	*Digitosporium*	**E4**
165(163)	Pseudopycnidium unilocular		166
	Pseudopycnidium multilocular with		
	complete or incomplete locules		169
166(165)	Conidia hyaline		167
	Conidia brown		168
167(166)	Conidia guttulate	*Rabenhorstia*	**E5**
	Conidia non-guttulate	*Hysterodiscula*	**E7**
168(166)	Conidia oval, ellipsoid or reniform,		
	often with the residues of the		
	conidiogenous cell at the base	*Harknessia*	**E3**
	Conidia reniform without basal fringe,		
	pseudopycnidium with clypeus	*Cymbothyrium*	**E8**
169(165)	Conidia hyaline or light-coloured		170
	Conidia brown		179
170(169)	Conidia hyaline		171
	Conidia pale brown		178
171(170)	Conidia unicellular		172
	Conidia bi- or multicellular		174
172(171)	Conidia more or less cylindrical or ovoid		173
	Conidia fusiform without basal fringe	*Fusicoccum*	**E10**
173(172)	Conidia ovoid, conidioma unilocular	*Polymorphum*	**E9**
	Conidia cylindrical, conidiomata in		
	columns joined at the base	*Crandallia*	(VI G14)
174(171)	Conidia bicellular	*Microperella*	(V C19)
	Conidia multicellular		175
175(174)	Conidiophores surrounded by a		
	mucilaginous sheath	*Placothyrium*	**E14**
	No such character		176
176(175)	Conidia with a subapical bristle at		
	each end	*Discosia*	**E18**
	Conidia muticate		177
177(176)	Phragmospores with blunt ends, fruit		
	body on conifer needles	*Dothistroma**	**E15**
	Scolecospores with pointed ends	*Pleurothyrium*	**E17**
178(170)	Conidia with a subapical bristle at		
	each end	*Discosia*	**E18**
	Conidia with a gelatinous apical		
	appendage	*Septoriella*	**E19**

---

*A closely related genus, *Pittostroma* (teleomorph = *Grovesiella*) has been described as being on fir branches, cf. *Mycologia* 85, 653-659, 1993.

179(169) Conidia unicellular 180
Conidia septate 181
180(179) Conidia smooth with germ slit *Endomelanconium* **E11**
Conidia spiny without germ slit *Aplosporella* **E12**
181(179) Dictyospores *Dichomera* **E20**
Didymo- or phragmospores 182
182(181) Conidia bicellular, presence of paraphyses *Lasiodiplodia* **E13**
Conidia multicellular without paraphyses *Neohendersonia* (V C14)

V

ANNELLOPHORAE AND ANNELLOBLASTOSPORAE
(Holoblastic Annellosporae)

We have seen earlier that this group comprises fungi that are basically similar to the Aleuriosporae and Monoblastosporae (chapter IV), but in which, typically, there is a percurrent proliferation of the conidiogenous cell. This results in a superposition of circular scars, giving it a characteristic ringed appearance that gives the group its name (Hughes, 1953). The scars are generally fairly elongated because the regrowth is significant. We have seen in the introduction (chapter I) that conidiogenous cells of a phialide appearance may also present this phenomenon of percurrent growth, which is why we divide this group of Annellosporae of Hughes into two. This corresponds to the distinction introduced by Wang (1990) between **annellophore**, conidiophore with intrahyphal growth and holoblastic conidia, and **annellide**, phialide with percurrent growth and enteroblastic conidia.

Conidial secession following growth seems to involve the lower part of the conidial wall in *Endophragmia, Phragmocephala,* and, to a lesser extent, *Melanocephala*. This results in a succession of scars in a funnel or cupule shape with very distinctive appearance, at the top of the conidiophore (Table V.B. 5, 6, and 11). *Phragmocephala,* the only genus with coremia, has been as an exception grouped with macronematous Hyphales (Table V.B 11), while the few genera with sporodochia are illustrated with the semi-macronematous Hyphales (Table V.A).

Generally upright, the conidiophores may be ramified (*Brachysporiella,* Table V.B 4; *Pseudocenangium,* Table V.D 25). In the Hyphales, they are almost entirely melanized and also most often macronematous.

The distinction between annellophores and annellides is nearly impossible to make under light microscope in the Annelloblastosporae.

Thus, in the genus *Coniothyrium* (Table V.C 10), considered holoblastic by Sutton (1980), we have found a species that has been noted to form phialides with sequential growth (Reisinger et al., 1977b). Its phialidic nature puts it in the genus *Microsphaeropsis* (Jones, 1976).

On the other hand, a certain number of genera, for example, *Sporidesmiella* (Table V.B 14), may grow percurrently or sympodially. *Leptographium* (Table VI.C 22), with percurrent growth, presents the appearance of sympodial growth when the conidia are laterally put out. Wingfield (1985) studied these phenomena in detail and reduced the genus *Verticicladiella* (sympodial) into synonymy with *Leptographium* (percurrent).

This is why a number of genera that belong to other groups, mainly the Annellidae (Chapter XI), will be found in the following key.

Key to identification of Annellophorae and Annelloblastosporae

1	Hyphales		2
	Conidiomales		38

Hyphales _____

2(1)	Conidiophores short, semi-macronematous		3
	Conidiophores long, macronematous		19

Semi-macronematous conidiophores _____

3(2)	Conidia surrounded by a covering that tears at the tip	*Conioscypha*	**A1**
	Typical annellophores*		4
4(3)	Ameroconidia		5
	Conidia septate		10
5(4)	Without synanamorph		6
	With synanamorph *Scopulariopsis*	*Wardomyces*	(IV A30)
6(5)	Cylindrical conidiogenous cell		7
	Bulging conidiogenous cell		8
7(6)	Conidia subcylindrical	*Belemnospora*	**A2**
	Conidia bulging	*Spilocaea*	**A8**
8(6)	Conidiogenous cell consisting of a bulging middle between a long foot and neck	*Torulomyces*	(X D3)
	Phialiform conidiogenous cell		9
9(8)	Conidia small with truncated base	*Scopulariopsis*	(XI A2)
	Conidia large, round with germ slit	*Wardomycopsis*	(XI A3)

_____

*See also *Molliardiomyces*, VI C 30, Sympodulosporae that can occasionally be annellated.

10(4)	Didymoconidia		11
	Phragmo- or dictyoconidia		13
11(10)	Conidia subcylindrical	*Belemnospora*	A2
	Conidia more or less bulging		12
12(11)	Short chains of conidia with round tip	*Bactrodesmiella*	A4
	Conidia with pointed tip	*Spilocaea*	A8
13(10)	Phragmoconidia		14
	Dictyoconidia		17
14(13)	Conidiophores short, isolated on a mycelium with hyphopodia	*Clasterosporium*	A9
	Conidiophores grouped more or less densely in natural conditions		15
15(14)	Conidiophores short on an intracellular mycelium, phragmoconidia with septal constrictions	*Mastigosporium*	A5
	Conidiophores erect in sporodochia		16
16(15)	Conidia short and pointed, rarely tricellular	*Spilocaea*	A8
	Conidia with more numerous septa	*Stigmina*	A7
17(13)	Small, muriform dictyoconidia		18
	Conidia large, in horseshoe form	*Eversia*	A6
18(17)	Conidiophores short, isolated	*Annellophorella*	A3
	Conidiophores erect in sporodochia	*Stigmina*	A7

Macronematous conidiophores*

19(2)	Conidia grouped in chains or heads		20
	Conidia isolated		22
20(19)	Conidia in mucous heads on verticillate conidiogenous cells	*Leptographium*	(VI C22)
	Conidia in chains		21
21(20)	Chain formed by acropetal budding (cf. Acroblastosporae)	certain *Endophragmia*	B5
	Conidia forming apical rings	*Annellophora*	B2
22(19)	Ameroconidia		23
	Conidia septate		24
23(22)	Conidiophores thin, conidia with basal fringe	*Domingoella*	B3
	Conidiophores larger, conidia without basal fringe	*Acrogenospora*	B9
24(22)	Conidia septate, without protuberances		25
	Conidia with apical or lateral protuberances, or staurospores		33
25(24)	Conidiophores branched (see also certain *Endophragmiella*)	*Brachysporiella*	B4

*See also *Cacumisporium* (VI D 6), Sympodulosporae that may occasionally be annellated.

	Conidiophores generally simple		26
26(25)	More or less cylindrical growths		27
	Bulging or funnel-shaped growths		
	(see also *Phragmocephala*)		30
27(26)	Rhexolytic secession	*Endophragmiella*	**B10**
	Schizolytic secession		28
28(27)	Phragmoconidia		29
	Dictyoconidia	*Acrodyctis*	**B18**
29(28)	Conidia elongated more or less		
	narrowed at the tip	*Sporidesmium*	**B8**
	Conidia rounded at the tip (sometimes		
	sympodial growths)	*Sporidesmiella*	**B14**
30(26)	Bulging proliferation		31
	Funnel-shaped proliferation (= cupule)		32
31(30)	Bulging, evenly coloured proliferation	*Sporidesmium*	**B8**
	Bulging proliferation darker at the base	*Deightoniella*	**B12**
32(20)	Cupule as large as the conidium (see		
	also *Phragmocephala*, with coremia)	*Endophragmia*	**B5**
	Cupule narrower than the conidium	*Melanocephala*	**B6**
33(24)	Staurospores	*Triposporium*	**B13**
	Conidia with protuberances or various		
	appendages		34
34(33)	Conidia elongated, pointed		35
	Conidia globular or ovoid		36
35(34)	Basal expansions	*Teratosperma*	**B7**
	Apical and lateral bristles	*Chaetendophragmia*	**B1**
36(34)	Acellular protuberances (sometimes		
	also an apical bristle)	*Acrophragmis*	**B15**
	Cellular protuberances		37
37(36)	Aseptate protuberances	*Uberispora*	**B16**
	Septate protuberances	*Arachnophora*	**B17**

Conidiomales

| 38(1) | Coremia | 39 |
| | Other conidiomata | 44 |

Coremia

39(38)	Ameroconidia		40
	Phragmoconidia		42
40(39)	Conidia mucous in a mass at tip of		
	coremia	*Graphium*	(VI I 2, XI, A12)
	Conidia dry (= *Scopulariopsis* with coremia)		41
41(40)	Coremia without bristles	*Doratomyces*	(XI A7)
	Coremia with sterile bristles	*Trichurus*	(XI A8)
42(39)	Presence of cupules on the		
	conidiogenous cells	*Phragmocephala*	**B11**
	No funnel-shaped cupules		43

98   The Deuteromycetes: Mitosporic Fungi

43(42)	Conidia generally triseptate at rounded		
	tip	*Arthrobotryum*	(XI A14)
	Conidia with tapering tip, percurrent		
	and/or sympodial growth	*Pseudocercospora*	(VI I 10)
44(38)	Pycnidia (see also 101 = some Annellidae)		45
	Other conidiomata		54

Pycnidia

45(44)	Conidia multicellular		46
	Conidia uni- or bicellular		49
46(45)	Conidia distoseptate, surrounded with		
	a mucous sheath	*Macrodiplodiopsis*	**C1**
	Conidia euseptate, without mucous sheath		47
47(46)	Conidia muriform	*Camarosporium*	**C2**
	Conidia non-muriform		48
48(47)	Conidia with cellular appendages	*Bartalinia*	**C3**
	Conidia without appendages	*Stagonospora*	**C4**
49(45)	Conidia hyaline		50
	Conidia more or less melanized		51
50(49)	Conidia muticate, curved	*Pseudoseptoria*	**C5**
	Conidia upright with apical bristles	*Neoalpakesa*	**C6**
51(49)	Ameroconidia		52
	Didymoconidia		53
52(51)	Generally lichenicolous fungi, echinulate		
	conidia	*Lichenoconium*	**C7**
	Fungi on higher plants, smooth conidia	*Sphaeropsis*	**C9**
53(51)	Cylindrical conidiogenous cell	*Schwarzmannia*	**C8**
	Bulging conidiogenous cell	*Coniothyrium*	**C10**
54(44)	Pseudopycnidia (see also 104 = some Annellidae)		55
	Sporodochia, acervuli or cupules		65

Pseudopycnidia

55(54)	Complex pseudopycnidia in multilocular crust		56
	Simple or multilocular pseudopycnidia		57
56(55)	Ameroconidia	*Lasmeniella*	**C16**
	Conidia (0)-1-(2) septate	*Dothideodiplodia*	**C15**
57(55)	Ostiolar dehiscence		58
	Dehiscence by rupture of the wall		63
58(57)	Conidia with appendages		59
	Conidia without appendages		60
59(58)	Cellular apical appendage, conidia		
	non-muriform	*Uniseta*	**C11**
	Mucilaginous basal appendage,		
	conidia muriform	*Shearia*	**C12**
60(58)	Ameroconidia	*Fusicoccum*	**C13**
	Conidia septate		61
61(60)	Didymo- or phragmoconidia		62
	Dictyoconidia	*Dichomera*	(IV E20)

62(61)	Hyaline didymoconidia	*Scaphidium*	**C20**
	Discoloured phragmoconidia	*Neohendersonia*	**C14**
63(57)	Conidia unicellular, curved	*Disculina*	**C18**
	Conidia bicellular		64
64(63)	Conidiomata unilocular, conidia guttulate with gelatinous apical appendages	*Comatospora*	**C17**
	Conidiomata irregularly multilocular, conidia without appendages	*Microperella*	**C19**
65(64)	Conidiomata cupuloid or discoid		66
	Conidiomata sporodochial or acervular		69

Cupules or discs

66(65)	Conidia hyaline or light-coloured		67
	Conidia brown or multicoloured		68
67(66)	Conidia in the form of an X or Y	*Ypsilonia*	**D26**
	Conidia filiform, bicellular	*Pseudocenangium*	**D25**
68(66)	Conidia brown, aseptate, stromatic conidiomata closed at first	*Hymenopsis*	**D24**
	Conidia multicoloured, distoseptate, with appendages	*Pestalotia*	**D29**
69(65)	Sporodochia		70
	Acervuli		75

Sporodochia

70(69)	Conidia hyaline		71
	Conidia melanized		73
71(70)	Ameroconidia; conidiogenous cells: phialides and annellides; hyperparasite	*Tympanosporium*	(XI A9)
	Conidia septate		72
72(71)	Didymoconidia	*Gymnodochium*	(XI A10)
	Phragmoconidia	*Mastigosporium*	**A5**
73(70)	Short annellophores	*Pollaccia*	(IV C5)
	Long annellophores		74
74(73)	Conidia (0)-1(2) septate	*Spilocaea*	**A8**
	Phragmo- or dictyoconidia	*Stigmina*	**A7**

Acervuli

75(69)	Conidia hyaline or light-coloured		76
	Conidia brown or multicoloured		80
76(75)	Fusoid, upright or curved ameroconidia	*Fusamen*	(VI H10)
	Conidia septate		77
77(76)	Conidia bicellular		78
	Conidia multicellular		79
78(77)	Conidia guttulate with septal constriction	*Marssonina*	**D1**
	Conidia non-guttulate, curved, conidiogenous cell partly melanized	*Anaphysmene*	**D2**

79(77)	Conidia hyaline	*Phloeospora*	**D3**
	Conidia light brown	*Colletogloeum*	**D4**
80(75)	Conidia brown, evenly coloured		81
	Conidia multicoloured		93
81(80)	Conidia distoseptate		82
	Conidia unicellular or euseptate		83
82(81)	Conidia muriform, paraphyses present	*Stegonsporium*	**D5**
	Phragmoconidia, paraphyses absent	*Coryneum*	**D6**
83(81)	Conidia unicellular		84
	Conidia bi- to multicellular		88
84(83)	Paraphyses present	*Lamproconium*	**D7**
	No paraphyses		85
85(84)	Conidia warty	*Leptomelanconium*	**D8**
	Conidia smooth		86
86(85)	Conidia pyriform to obpyriform	*Melanconiopsis*	**D9**
	Conidia not as above		87
87(86)	Conidia cylindrical, cuneiform	*Piggotia*	**D10**
	Conidia globular to ellipsoid	*Melanconium*	**D11**
88(83)	Conidia bicellular		89
	Conidia multicellular		90
89(88)	Conidia smooth	*Didymosporina*	**D12**
	Conidia warty	*Leptomelanconium*	**D8**
90(88)	Paraphyses present	*Stilbospora*	**D13**
	No paraphyses		91
91(90)	Conidia filiform, fusoid, often curved	*Lecanosticta*	**D14**
	Conidia fusoid or clavate (club-shaped)		92
92(91)	Conidia club-shaped	*Gloeocoryneum*	**D15**
	Conidia ovoid, narrow between septa	*Sporocadus*	**D17**
93(80)	Dictyoconidia	*Morinia*	**D18**
	Phragmoconidia		94
94(93)	Conidia with pointed tips, but no appendage	*Toxosporium*	**D16**
	Conidia with appendages		95
95(94)	Exogenous (eccentric) basal appendage		96
	Endogenous (central) basal appendage or no basal appendage		98
96(95)	Conidia 3-septate and with two appendages at each end	*Diploceras*	**D19**
	Conidia with a single basal appendage sometimes ramified		97
97(96)	Conidia 2- to 3-septate, apical appendage absent or present, sometimes ramified	*Seimatosporium*	**D20**
	Conidia 4- to 5-septate, simple or double apical appendage	*Sarcostroma*	**D21**
98(95)	Conidia with an apical appendage sometimes ramified		99
	Conidia with several apical appendages		100

Table V.A. Semi-macronematous Hyphales

Fig.	Genus	No. spp.	Geographical distribution	Mode of life, substrates and impact	Teleomorphs	Additional references	Mol. biol.
1	*Conioscypha*	4	Cosmopolitan	Saprophyte on wood and bark			
2	*Belemnospora*	4	Ubiquitous	Saprophyte on litter		Myc. Res. 92: 354 (1989)	
3	*Annellophorella*	4	Cosmopolitan	Parasite and saprophyte on plants			
4	*Bactrodesmiella*	2	Gr. Britain, New Zealand	Saprophyte on plants			
5	*Mastigosporium*	5	Eurasia	Parasite on plants			
6	*Eversia*	1	Temperate North	Saprophyte on wood			
7	*Stigmina*	58	Cosmopolitan	Parasite and saprophyte on plants	*Acantharia, Gibbera, Mycosphaerella, Otthia*		
8	*Spilocaea*	6	Cosmopolitan	Parasite on plants	*Venturia*		X
9	*Clasterosporium*	10	Cosmopolitan	Parasite on plants	*Asterina, Asterodothis*		

Semi-macronematous Hyphales

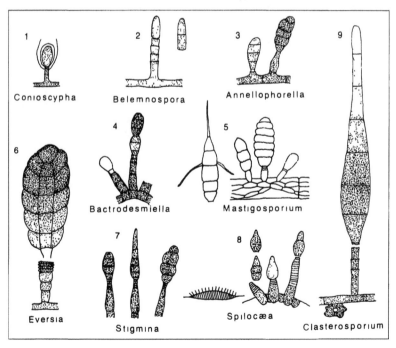

Conioscypha
Belemnospora
Annellophorella
Bactrodesmiella
Mastigosporium
Eversia
Stigmina
Spilocæa
Clasterosporium

Table V.B. Macronematous Hyphales

Fig.	Genus	No. spp.	Geographical distribution	Mode of life, substrates and impact	Teleomorphs	Additional references	Mol. biol.
1	*Chaetendophragmia*	5	Cosmopolitan	Saprophyte on plants			
2	*Annellophora*	7	Cosmopolitan	Saprophyte on plants			
3	*Domingoella*	4	Gr. Britain, tropical zone	Saprophyte on plants or hyperparasite			
4	*Brachysporiella*	4	Ubiquitous	Saprophyte on plants or aquatic	*Ascotaiwania*		
5	*Endophragmia*	7	Temperate zone	Saprophyte on plants			
6	*Melanocephala*	3	Oceania, USA	Saprophyte on plants			
7	*Teratosperma*	7	Ubiquitous	On plants, lichens and fungi			
8	*Sporidesmium*	± 70	Cosmopolitan	Mostly saprophyte on plants	*Alina, Eriosphaeria, Eupelte*		
9	*Acrogenospora*	6	Cosmopolitan	Saprophyte on plants	*Farlowiella*	Myc. Res. 102: 1309 (1998)	
10	*Endophragmiella*	48	Cosmopolitan	Saprophyte on plants	*Lasiosphaeria*	N.Z.J. Bot. 17: 139: Mycot 23: 305 (1985)	
11	*Phragmocephala*	5	Temperate North, Cuba	Saprophyte on plants			
12	*Deightoniella*	9	Cosmopolitan mostly tropical	Saprophyte and 1 parasite on plants			
13	*Triposporium*	2	Ubiquitous	Saprophyte and parasite on plants	*Batistinula*		
14	*Sporidesmiella*	18	Cosmopolitan	Saprophyte on plants		Myc. Res. 102: 548 (1998)	
15	*Acrophragmis*	3	Ubiquitous	Saprophyte on wood			
16	*Uberispora*	2	India, Japan, USA	Saprophyte on plants			
17	*Arachnophora*	2	Temperate North	Saprophyte on plants			
18	*Acrodictys*	20	Cosmopolitan, mostly tropical	On plants and fungi			

Macronematous Hyphales

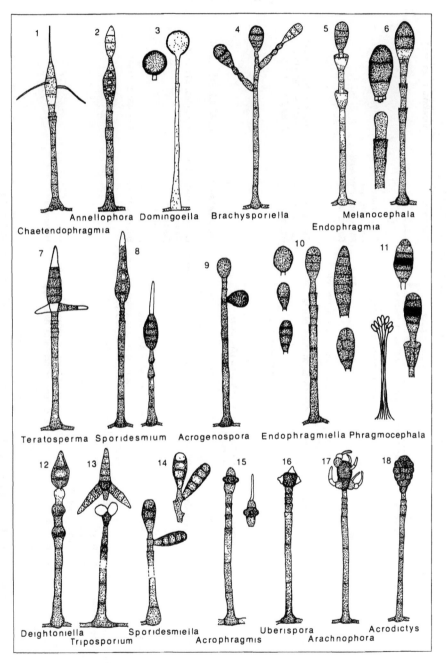

Chaetendophragmia — Annellophora — Domingoella — Brachysporiella — Melanocephala — Endophragmia

Teratosperma — Sporidesmium — Acrogenospora — Endophragmiella — Phragmocephala

Deightoniella — Triposporium — Sporidesmiella — Acrophragmis — Uberispora — Arachnophora — Acrodictys

Table V.C. Annelloblastosporae with pycnidia or pseudopycnidia

Fig.	Genus	No. spp.	Geographical distribution	Mode of life, substrates and impact	Teleomorphs	Additional references	Mol. biol.
1	Macrodiplodiopsis	1	N. America. Europe	Saprophyte on dead branches of Platanus			
2	Camarosporium	100	Ubiquitous, mostly temperate North	Saprophyte on leaves, twigs, stem and branches of many plants	Cucurbitaria, Leptosphaeria, Pleospora, Pleoseptum, Splanchnonema		
3	Bartalinia*	7	Cosmopolitan	Saprophyte or parasite on various organs of woody plants			
4	Stagonospora	200	Cosmopolitan	Parasite or saprophyte on many plants, 77 species graminicolous	Didymella, Leptosphaeria, Pleospora	O.U.P. (1965, 1967); TBMS 50: 85 (1967); AFSAT 10: 1 (1977)	X
5	Pseudoseptoria	5	Ubiquitous	Agent of leaf necrosis in Poaceae			
6	Neoalpakesa	1	Gr. Britain	On leaves of Poa glauca			
7	Lichenoconium	10	Ubiquitous	Lichenicolous hyperparasite, 1 sp. on Pinus wood			
8	Schwarzmannia	1	Russia	Parasite on living leaves (Goebelia, Ammodendron)			
9	Sphaeropsis	30	Cosmopolitan	Parasite (canker, dieback) or saprophyte on various organs of many plants, endophyte	Botryosphaeria	EJFP 23: 317 (1993)	X
10	Coniothyrium	26	Cosmopolitan	On various organs of many plants, 1 sp. parasite on rose plants	Curreya, Massarina, Paraphaeosphaeria, Pleospora	Myc. Res. 97: 1175 (1993)	X
11	Uniseta	1	USA	On branches of Comptonia	Cryptodiaporthe		
12	Shearia	2	N. Amer.	On branches and wood of Magnolia	Pleomassaria		
13	Fusicoccum	± 50	Cosmopolitan	Parasite (canker, dieback) or saprophyte on various organs of many plants, endophyte	Botryosphaeria	RFF 45: 37 (1993)	
14	Neohendersonia	1	N. Amer., Europe	On branches and bark of Fagus sylvatica			
15	Dothideodiplodia	1	Russia	Saprophyte on dry culms of Agropyron repens			
16	Lasmeniella	13	Ubiquitous, mostly tropical	On living or dead leaves of many plants			
17	Comatospora	1	Canada	Saprophyte on dead twigs of Picea mariana			
18	Disculina	1	Europe	On twigs and dead branches of Alnus spp.	Winterella		
19	Microperella	1	Japan	Parasite on leaves of Quercus glauca			
20	Scaphidium	1	USA	Agent of leaf lesions of Bouteloua			

*A reexamination of the species of this genus has shown that they are not annellidate (Nag Raj, 1993).

Annelloblastosporae with pycnidia or pseudopycoidia

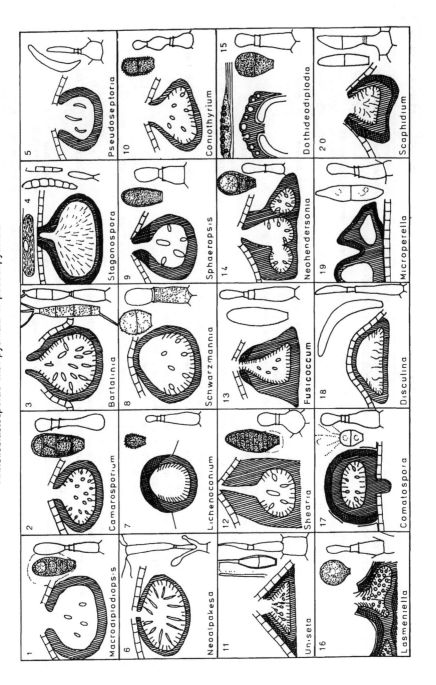

Table V.D. Annelloblastoporae with acervuli and cupules

Fig.	Genus	No. spp.	Geographical distribution	Mode of life, substrates and impact	Teleomorphs	Additional references	Mol. biol.
1	Marssonina	70	Ubiquitous, mostly temperate zone	Chiefly leaf parasite, maculicolous. on many woody or herbaceous plants	Diplocarpon, Drepanopeziza	Mycol. 85: 814 (1983)	
2	Anaphysmene	2	Europe, Guatemala	On stem of Heracleum sphondylium and dead leaves of Cupressus		Mycol. 82: 313 (1990)	
3	Phloeospora	160	Ubiquitous	Generally a leaf parasite, maculicolous on many mostly woody plants	Mycosphaerella		
4	Colletogloeum	9	Cosmopolitan	Leaf parasite, maculicolous	Mycosphaerella	TBMS 87: 93 (1986)	
5	Stegonsporium	2	N. Amer., Europe	On twigs and bark of trees (Acer, Aesculus, Fagus, Tilia)	Splanchnonema	Myc. Pap. 145 (1981)	
6	Coryneum	21	Ubiquitous, mostly temperate	On twigs, branches, bark of trees (Quercus, Betula, Carpinus, Alnus, Castanea, Aesculus, Ulmus...). endophyte	Pseudovalsa	N. Hedw. 32: 341 (1980)	
7	Lamproconium	1	Europe	On branches of lime (Tilia spp.)			
8	Leptomelanconium	3	N. Amer., Australia, Europe	On needles (Pinus spp.), cones (Picea), leaf lesions (Eucalyptus)			
9	Melanconiopsis	2	N. Amer., Europe	On branches of trees (Acer...)	Massariovalsa		
10	Piggotia	2	Temperate zone	Leaf parasite (Corylus, Ulmus)	Platychora		
11	Melanconium	50	Ubiquitous	On bark and branches of various trees	Melanconis	CJB 69: 2170 (1991)	
12	Didymosporina	1	Europe	Agent of leaf lesions on maples (Acer spp.)			
13	Stilbospora	15	Ubiquitous	On branches and leaves of many woody plants			
14	Lecanosticta	4	Ubiquitous	Maculicolous pathogen of needles of pine (Pinus spp.)			
15	Gloeocoryneum	1	USA	On needles of Pinus	Eruptio	Myc. Pap. 153	X
16	Toxosporium	2	N. Amer., Europe	On needles of conifers and wood of poplar			
17	Sporocadus	16	Ubiquitous	On various organs of trees and shrubs	Discostroma		
18	Morinia	1	Eurasia	On dead branches of Artemisia			
19	Diploceras	1	Canada, Europe	On stems of Hypericum spp.	Saccothecium		
20	Seimatosporium	18	Cosmopolitan	On various living or dead organs of trees and shrubs, sometimes maculicolous	Discostroma		
21	Sarcostroma	20	Cosmopolitan	On various organs of many plants			
22	Monochaetia	17	Cosmopolitan	On various dead or moribund organs of trees and shrubs			
23	Pestalotiopsis	85	Cosmopolitan	On various organs of many plants	Pestalosphaeria	T.M.S. Jap. 34: 323 (1993)	X
24	Hymenopsis	7	Ubiquitous	On leaves and stems of Cyperaceae, Poaceae and Fabaceae			
25	Pseudocenangium	2	N. Amer., Europe	On needles of Pinus spp.	? Cyathicula		
26	Ypsilonia	8	Ubiquitous, mostly tropical	On scales of dead insects, leaves, rotting bark	Acanthotheciella		
27	Seiridium	22	Ubiquitous	On various organs of many plants, pathogen on Cuppressaceae (agent of canker)	Blogiascospora, Lepteutypa	Syd. 38: 179 (1986); EJFP 19: 435 (1989)	X
28	Truncatella	5	Ubiquitous	On various living or dead orgns of many woody or herbaceous plants	Broomella		
29	Pestalotia	1	Europe, USA	Branches and twigs of Vitis			

Annelloblastosporae with acervuli and cupules

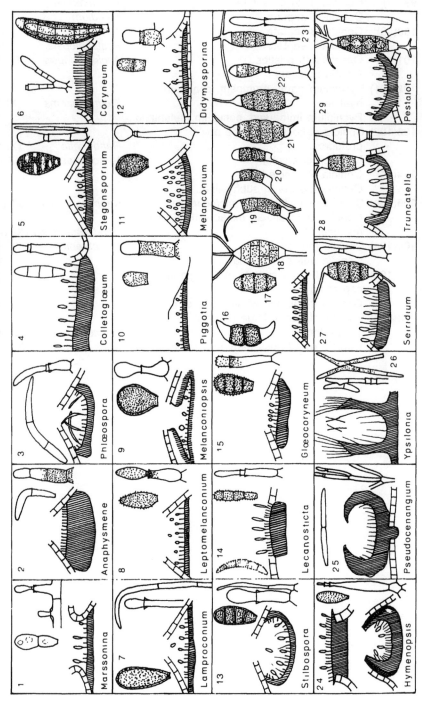

1 Marssonina. 2 Anaphysmene. 3 Phlœospora. 4 Colletoglœum. 5 Stegonsporium. 6 Coryneum. 7 Lamproconium. 8 Leptomelanconium. 9 Melanconiopsis. 10 Piggotia. 11 Melanconium. 12 Didymosporina. 13 Stilbospora. 14 Lecanosticta. 15 Gloeocoryneum. 16 Toxosporium. 17 Sporocadus. 18 Morinia. 19 Diploceras. 20 Seimatosporium. 21 Sarcostroma. 22 Monochaetia. 23 Pestalotiopsis. 24 Hymenopsis. 25 Pseudocenangium. 26 Ypsilonia. 27 Seiridium. 28 Truncatella. 29 Pestalotia.

99(98)	Conidia 3- to 4-septate	*Monochaetia*	**D22**
	Conidia 5-septate	*Seiridium*	**D27**
100(98)	Conidia 3-septate	*Truncatella*	**D28**
	Conidia 4-septate	*Pestalotiopsis*	**D23**
101(44)	Ostiole with bristles	*Angiopomopsis*	(XI B7)
	Ostiole muticate		102
102(101)	Pycnidium partly erumpent, conidium hyaline	*Trematophoma*	(XI B5)
	Pycnidium immersed, conidium melanized		103
103(102)	Conidia deltoid with three apical protuberances	*Readeriella*	(XI B4)
	No such character	*Microsphaeropsis*	(X O9)
104(54)	Conidia melanized		105
	Conidia hyaline to subhyaline		106
105(104)	Conidia unicellular, no microconidia	*Neomelanconium*	(XI B14)
	Conidia multicellular, mixed with hyaline microconidia	*Hendersoniopsis*	(XI B13)
106(104)	Conidia multicellular	*Phlyctaeniella*	(XI B8)
	Conidia uni- or bicellular		107
107(106)	Conidia 0- to 1-septate, mucilaginous cap at tip	*Sphaerellopsis*	(XI B12)
	Conidia 0-septate, without cap at tip		108
108(107)	Conidia ellipsoid, large (more than 13 μm long)	*Cryptosporiopsis*	(XI B2)
	Conidia ovoid, less than 13 μm long	*Discosporium*	(XI B10)

VI

SYMPODULOSPORAE

The term 'sympodulospores' designates conidia inserted on the conidiophore by the intermediary of a more or less narrow isthmus (Fig. VI.1). At maturity, when the conidium separates, this isthmus breaks off and leaves:

— a hilum at the base of the conidium: the conidia are apiculate (pointed hilum), truncated (flat hilum), or scarred (hilum rounded, more or less conforming to the general outline of the conidium);

— a more or less pointed and persistent conidial scar on the conidiophore, with forms similar to those of the hilum.

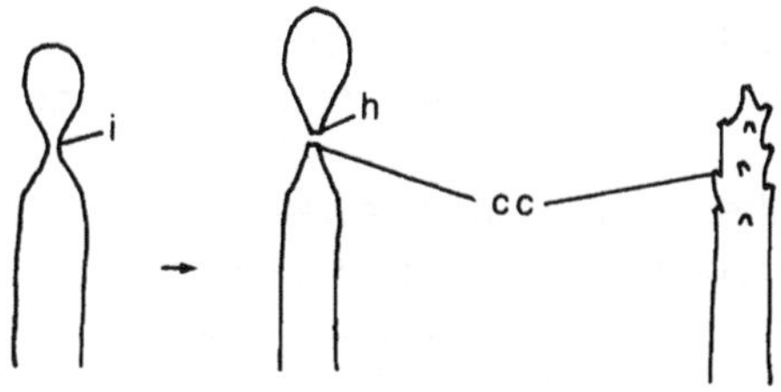

Fig. VI.1. Formation of conidia in the Sympodulosporae.
I: isthmus of insertion of the conidia on the conidiophore; h: hilum at the base of the conidium; cc: conidial scar on the conidiophore.

The production of conidia and the development of the sporogenous cell are depicted in Fig. VI.2.

The conidial primordium is a bud that appears generally at the tip of the conidiophore. The conidium has a wall continuous with that of the conidiogenous cell, and it is therefore **holoblastic**. It remains single, in the sense that:

— no other conidium can be produced on the same vegetative point;

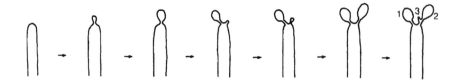

Fig. VI.2. Development of the sporogenous cell of Sympodulosporae up to the production of conidia.

— the conidium does not itself bud into other conidia.

The conidiogenous cell puts out a short growth, from a subapical point, and its fertile tip gives rise to the next conidium. Depending on the position of fertile growths, their length, etc., the result is two characteristic appearances:

— the grated appearance of the conidiophore seen in Fig. VI.1 (e.g., *Virgaria*, Table VI.C 14), from which the name of Radulasporae was given to this group by Mason (1933, from the Latin *radula*, grater, scraper);

— the appearance of a zigzag rachis seen in Fig. VI.2 (e.g., *Beauveria*, Table VI.C 9). It is this aspect of the sympodial growth that gives the group its present name (Barron, 1968).

However, we must note that:

— this sympodial growth is found in other groups (e.g., *Curvularia*, Porosporae, Table IX.17; *Sympodiella*, Arthrosporae, Table II.A 16; *Ramularia*, Acroblastosporae, Table VII.A 7);

— the radula or the sympode may be more or less scarred and therefore much less distinct, the conidial scars being reduced to a nearly invisible point.

Moreover, all the intermediates between these two types (radula and zigzag rachis) are found and the former may be reduced to a short, rather bulging sporogenous head (e.g., *Nodulisporium*, Table VI.C 4; *Pseudobotrytis*, Table VI.E 4). In any case, the conidia are terminal at the origin and do not become lateral till after the growth of the sporogenous zone; one can therefore describe them as pseudopleurogenous (they appear to be produced laterally, but they are actually acrogenous).

The conidiophore may be more or less developed and present all stages between a simple fertile ramification of the vegetative hypha and a well-differentiated sporiferous apparatus. In some genera (*Arthrobotrys*, Table VI.B 2), it proliferates and carries fertile nodes at different heights.

The conidia are separated by various mechanisms. The often reduced size of the spicules does not always allow us to detect the mode of separation under light microscope. In the case of *Conoplea* (Table VI.C 27), the isthmus is large and hollow (disjunctor). It bursts in the middle, between the two plugs that block off the conidium on the one hand and the sporogenous cell on the other hand (Reisinger, 1966). In *Acrostaphylus* (Table VI.C 15) and *Beauveria* (Table VI.C 9), this hollow part does not exist but is replaced by a significant thickness of wall (Fig. VI.3).

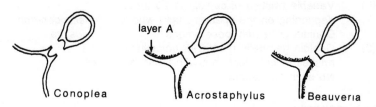

layer A
Conoplea   Acrostaphylus   Beauveria

Fig. VI.3. Different mechanisms of conidial separation in the Sympodulosporae, revealed under electron microscope. In *Conoplea*, the large and hollow isthmus bursts in the middle (rhexolytic secession at the disjunctor). In *Acrostaphylus* and *Beauveria*, the hollow part does not exist, and the secession is schizolytic.

In *Brachysporium* (Table VI.D 10), for example, the isthmus is long and probably functions as in *Conoplea*.

Certain genera may have percurrent proliferation in addition to sympodial growth, for example, *Leptographium* (Table VI.C 22) or *Sporidesmiella* (Table VI.D 13).

Key to identification of Sympodulosporae

1	Hyphales		2
	Conidiomales		118

Hyphales _____

2(1)	Conidia twisted in a helix or spiral		3
	Conidia not twisted		11

Helicospores _____

3(2)	Conidia in spiral (twisted in one plane)		4
	Conidia in helix (twisted in space)		7
4(3)	Conidial apparatus hyaline	*Helicomyces*	A1
	Conidial apparatus melanized		5
5(4)	Single conidia		6
	Conidia in chains	*Helicodendron*	A6
6(5)	Conidia hyaline or subhyaline, narrow	*Helicosporium*	A2
	Conidia hyaline to melanized, wide	*Helicoma*	A3
7(3)	Conidia more or less erect, curved, or slightly twisted	*Helicomina*	A4

	More significant twisting, general form of a cylinder or keg		8
8(7)	Single conidia		9
	Conidia in chains		10
9(8)	Conidial filament of constant size	*Helicoon*	**A5**
	Conidia narrow at the septa	*Helicorhoidion*	**A8**
10(8)	Variable twisting of conidia (1-20 turns depending on the species), tight whorls	*Helicodendron*	**A6**
	Conidia wide with loose whorls	*Hiospira*	**A7**
11(2)	Conidia generally rhomboid, unicellular, bristles or conidiophores with lobed basal cell		12
	No such characters		18

Rhombospores

12(11)	Conidiophores grouped in coremium	*Rhombostilbella*	**A15, I 7**
	Conidiophores separate		13
13(12)	Conidiophores setiform with fertile cells at different heights		14
	No such character		15
14(13)	Conidia biconic	*Beltraniopsis*	**A10**
	Conidia turbinate (like ninepin or top)	*Beltraniella*	**A11**
15(13)	Conidiophores and sterile bristles		16
	No sterile bristles		17
16(15)	Conidia biconic with apical appendage	*Beltrania*	**A9**
	Conidia turbinate	*Ellisiopsis*	**A12**
17(15)	Conidia biconic	*Pseudobeltrania*	**A13**
	Conidia obovoid	*Hemibeltrania*	**A14**
18(11)	Fertile part of the conidiogenous cell typically grated, elongated or subspherical, or zigzag rachis = Sympodulosporae s.s.		19
	No such characters		62

Sympodulosporae Hyphales s.s.

19(18)	Sporiferous apparatus hyaline		20
	Sporiferous apparatus melanized		36
20(19)	Fertile zone more or less isodiametric		21
	Fertile zone more or less elongated and grated or zigzag		23
21(20)	No marked conidial scars, nor well-differentiated conidiogenous cells	*Molliardiomyces*	**C30**
	Well-differentiated conidiogenous cells with clearly marked conidial scars		22
22(21)	Conidia hyaline, fertile zone in the form of a cock's comb	*Costantinella*	**C7**
	Conidia more or less melanized	*Basifimbria*	**C8**
23(20)	Rasped fertile zone		24
	Zigzag fertile zone		35

24(23)	Conidiophores hardly or not ramified		25
	Conidiophores more or less ramified		31
25(24)	Amerospores		26
	Didymospores	*Veronaea*	**D4**
26(25)	Integrated scars, spores pyriform, cuneiform (and aleuriospores)	*Raffaelea*	**C11**
	Prominent scars		27
27(26)	A single fertile terminal cell (a cell that can proliferate)		28
	Sporogenous zone divided into several cells by transverse septa	*Acladium*	**C20**
28(27)	Conidia globular	*Rhinotrichella*	**C10**
	Conidia more or less ovoid or elongated		29
29(28)	Small sympodulospores and large aleuriospores	*Disporotrichum*	(IV A13)
	No such character		30
30(29)	Conidia and scars small	*Sporothrix*	**C3**
	Conidia and scars larger	*Calcarisporiella*	**C5**
31(24)	Branches irregular	*Nodulisporium*	**C4**
	Branches verticillate		32
32(31)	Branches in cross formation, by threes		34
	Branches of variable number		33
33(32)	Amerospores	*Calcarisporium*	**C6**
	Phragmospores	*Arnoldiomyces*	**E1**
34(32)	Amerospores	*Geniculodendron*	**C2**
	Phragmospores	*Pseudohansfordia*	**E2**
35(23)	Well-developed rachis on a swollen base	*Beauveria*	**C9**
	Very well-developed rachis, branched conidiophore	*Lomentospora*	**C1**
36(19)	Grated fertile zone		37
	Zigzag fertile zone		56
37(36)	Fertile cells phialiform, grouped around sterile bristles or on setiform conidiophores Conidia allantoid		38
	No such characters		40
38(37)	Fertile cells grouped in a muff around the base of a ramified conidiophore	*Ceratocladium*	**F1**
	Fertile cells on the substrate		39
39(38)	Bristles circinate	*Circinotrichum*	**F2**
	Bristles ramified	*Gyrothrix*	**F3**
40(37)	Conidiophores hardly or not ramified		41
	Conidiophores more or less ramified		45
41(40)	Conidia unicellular		42
	Conidia bicellular	*Veronaea*	**D4**
42(41)	Conidiophores with poorly marked scars		43
	Conidiophores with pointed scars		44
43(42)	Conidia small and numerous	*Leptodontidium*	**C12**
	Conidia large and few	*Virgariella*	**C19**

44(42)	Conidia ellipsoid or elongated	*Rhinocladiella*	**C13**
	Conidia fusiform, curved	*Idriella*	**C21**
45(40)	Fertile zone elongated		46
	Fertile zone subspherical		52
46(45)	Disjunctors present	*Hansfordia*	**C25**
	Disjunctors absent		47
47(46)	Amerospores		48
	Phragmospores	*Dactylium*	**E3**
48(47)	Sporogenous cells swollen at base		49
	Sporogenous cells more or less cylindrical		50
49(48)	Conidia globular	*Acrodontium*	**C18**
	Conidia elongated	*Selenosporella*	**C29**
50(48)	Fertile zone isodiametric or slightly elongated	*Acrostaphylus*	**C15**
	Fertile zone elongated		51
51(50)	Conidia hyaline	*Geniculosporium*	**C16**
	Conidia melanized	*Virgaria*	**C14**
52(45)	Branches verticillate, didymospores	*Pseudobotrytis*	**E4**
	No such characters		53
53(52)	Branches lateral, compact	*Kumanasamuha*	**F7**
	Branches more or less regular and elongated		54
54(53)	Conidia small, guttuliform		55
	Conidia large, subspherical	*Dicyma*	**C26**
55(54)	Conidial apparatus hyaline	*Nodulisporium*	**C4**
	Conidial apparatus melanized	*Acrostaphylus*	**C15**
56(36)	Large fertile zone, melanized, often twisted		57
	Fertile zone more narrow, hyaline		58
57(56)	Conidia on pedicels more or less long	*Conoplea*	**C27**
	Conidia sessile, annular scars	*Polythrincium*	(IX 28)
58(56)	Conidiophores generally not ramified	*Parasympodiella**	**D16**
	Conidiophores ramified		59
59(58)	Branches 2- to 3-verticillate		60
	Branches non-verticillate		61
60(59)	Branches in arbuscule ending in sterile bristles	*Puciola*	**C24**
	No such character	*Tritirachium*	**C17**
61(59)	Converging penicilli, compressed conidiogenous cells	*Leptographium*	**C22**
	Diverging penicilli, metula almost perpendicular to the axis	*Verticicladium*	**C23**
62(18)	Conidiophores hyaline or subhyaline		63
	Conidiophores melanized		83

---

*See also *Sympodiella* (II A 16)

| 63(62) | Conidia single or carried on fertile nodes with blunt or pedicellate scars; nematophagous fungi | | 64 |
| | Non-nematophagous fungi of different types | | 72 |

Nematophagous fungi

64(63)	Staurospores	*Tridentaria*	**B9**
	Other forms of spores		65
65(64)	Amerospores, clamped hyphae, fusoid conidia on spicules	*Nematoctonus*	**B7**
	Conidia septate		66
66(65)	Didymospores		67
	Phragmospores		70
67(66)	Conidia cylindrical or conical, carried on long pedicels	*Candelabrella*	**B1**
	Bulging upper cell		68
68(67)	Fertile nodes with multiple spicules	*Arthrobotrys*	**B2**
	Conidiophores of different types		69
69(68)	Conidiophores geniculate	*Geniculifera*	**B6**
	No such character	*Duddingtonia*	**B8**
70(66)	Conidia subcylindrical	*Dactylaria*	**B3, D1**
	Conidia more or less bulging		71
71(70)	Conidia generally single, in a very swollen spindle	*Monacrosporium*	**B5**
	Conidia ellipsoid, fusoid or cylindrical, in small numbers	*Dactylella*	**B4**

Other Sympodulosporate Hyphales

72(63)	Conidiophores and conidia hyaline		73
	Conidiophores hyaline, conidia melanized		77
73(72)	Conidiophores ramified, conidia sometimes in chains with disjunctors	*Chrysosporium*	(IV A9)
		or *Geomyces*	(IV A12)
	No conidial chains		74
74(73)	Conidiophores extended by sterile bristles	*Botryoderma*	(IV A20)
	Conidiophores without sterile bristles		75
75(74)	No marked conidial scars, nor well-differentiated parent cells	*Molliardiomyces*	**C30**
	Conidiogenous cells and scars well-differentiated		76
76(75)	Conidiogenous cells in a cross	*Sporotrichum*	(IV A10)
	Short lateral branches with spiny conidia	*Histoplasma*	(IV A5)
77(72)	Conidia unicellular		78
	Conidia dictyate	*Oncopodiella*	(IV B8)
78(77)	Conidia globular		79
	Conidia elongated and more or less pointed		80

79(78)	Conidiophores and sterile bristles	*Botryotrichum*	(IV A41)
	⌐ No sterile bristles	*Gilmaniella*	(IV A33)
80(78)	Without annellosporate synanamorph	*Asteromyces*	(IV A32)
	With annellosporate synanamorph		81
81(80)	Conidiophores differentiated, regularly ramified, conidiogenous cells compact	*Wardomyces*	(IV A30)
	Conidiophores poorly differentiated		82
82(81)	Conidia smooth with germ slit	*Mammaria*	(IV A28)
	Conidia spiny without germ slit	*Echinobotryum*	(IV A29)
83(62)	Conidiophores macronematous, long, well-differentiated, with a bunch of conidia at the tip; generally lignicolous		84
	Conidiophores micronematous or semi-macronematous		106
84(83)	Branches present		85
	Branches absent		92
85(84)	Branches dichotomic or irregular		86
	Branches in a cross or penicillate		88
86(85)	Absence of conidiogenous cells or disjunctors	*Staphylotrichum*	(IV A40)
	Presence of conidiogenous cells or disjunctors		87
87(86)	A single conidium per conidiogenous cell	*Kramasamuha*	**F4**
	Two or several conidia per conidiogenous cell	*Balanium**	**F10**
88(85)	Branches in a cross	*Pleurothecium*	**D5**
	Branches penicillate apical		89
89(88)	Amerospores		90
	Phragmospores		91
90(89)	Converging brushes	*Leptographium*	**C22**
	Diverging brushes	*Verticicladium*	**C23**
91(89)	Conidia dry	*Parapyricularia*	**E5**
	Conidia mucous	*Sterigmatobotrys*	**E6**
92(84)	Fertile zone subspherical or elongated		93
	Fertile cells grouped on a terminal ampulla	*Xylocladium*	**F11**
93(92)	Fertile zone subspherical		94
	Fertile zone more or less elongated		97
94(93)	Verticillate fertile cells on the conidiophore	*Spondylocladiopsis*	**E7**
	Fertile zone apical		95
95(94)	No collarette	*Cordana*	**D2**
	A collarette, appearance of phialides		96

---

*Kirk (1985) created the genus *Balaniopsis* for a species with percurrent proliferation of conidiophore.

96(95)	Conidiogenous cell more or less cylindrical	*Cacumisporium*	**D6**
	Conidiogenous cell more or less bulging	*Blastophorum*	**D7**
97(93)	Disjunctor between conidiogenous cell and conidium	*Pyricularia*	**D11**
	No such character		98
98(97)	Blunt scars		99
	Wide and/or elongated scars, phragmospores		101
99(98)	Phragmospores	*Pleurophragmium*	**D8**
	Dictyospores		100
100(99)	Conidia medium-sized, continuous outline	*Dactylosporium*	**D9**
	Conidia large, with bulging cells	*Sirosporium*	**E8**
101(98)	Conidia cylindrical or with pointed tip		102
	Conidia pyriform, ovoid with rounded tip, sympodial or percurrent proliferation	*Sporidesmiella*	**D13**
102(101)	Cylindrical phragmoconidia, alternating along the length of the rachis	*Parasympodiella*	**D16**
	No such character		103
103(102)	Conidia fusoid without apical bristle	*Pseudospiropes*	(IX 9)
	Conidia with apical bristle(s)		104
104(103)	Conidia with tip tapering to a single bristle	*Subulispora*	**D15**
	Conidia with one or several distinct apical bristles		105
105(104)	Conidial scars long and narrow	*Camposporium*	**D17**
	Scars integrated, wide	*Pleiochaeta*	**D18**
106(83)	Conidiophores isolated		107
	Conidiophores in a bundle, often on a stroma		115
107(106)	Conidiophores micronematous		108
	Conidiophores semi-macronematous		111
108(107)	Bulging conidiogenous cells inserted on hyphae		109
	Conidiogenous cells inserted on arched or winding, melanized cells		110
109(108)	Phragmospores	*Polyschema*	**F5**
	Phragmospores and scolecospores	*Dwayabeeja*	**F6**
110(108)	A single amerospore per parent cell	*Zygosporium*	**F9**
	Several didymospores per parent cell	*Zygophiala*	**F8**
111(107)	Short or poorly branched conidiophores		112
	Conidiophores more or less long and indeterminate, fertile and more or less bulging cells		114
112(111)	Scars blunt		113

Scars pointed	*Scolecobasidium**	**D14**
113(112) Scars all around the conidiogenous cell	*Fusicladium*	**H12**
Alternating scars, staurospores	*Diplocladiella*	**F12**
114(111) Amerospores	*Paraphaeoisaria*	**C28**
Didymospores	*Scolecobasidiella*	**D3**
115(106) Conidial scars poorly marked	*Pseudocercosporella*	**E12**
Conidial scars thick, melanized		116
116(115) Very marked conidial scars on the bends of the conidiogenous cell, hilum thick and melanized	*Cercosporidium*	**E10**
Hilum truncated without noticeable thickness or melanization		117
117(116) Conidia (0)-1-(3) septate, basal cell swollen and apical cell elongated	*Passalora*	**E11**
Conidia elongated, multiseptate	*Cercospora*	**E9**

Conidiomales

118(1) Conidiomata pycnidial or pseudopycnidial		119
No such characters		144
119(118) Pycnidial conidiomata (cf. also *Polystigmina* G 17)		120
Pseudopycnidia conidiomata		126

Pycnidia

120(119) Pycnidium with one clypeus, septate filiform spores	*Eriospora*	**G1**
Pycnidium without clypeus		121
121(120) Parasite of Ascomycete and Ascolichen	*Cornutispora*	**G2**
Not on Ascomycete or Ascolichen		122
122(121) Conidium filiform, multicellular, simple or ramified		123
Conidium unicellular, not ramified		125
123(122) Conidium ramified with branches joined at a common base	*Dendroseptoria*	**G6**
Conidium non-ramified		124
124(123) Wall of pycnidium entirely of *textura angularis*	*Septoria*	**G3**
Pycnidium with lateral and basal wall of *textura oblita*	*Pocillopycnis*	**G23**
125(122) With ramified apical appendage	*Pestalozziella*	**G4**
Without apical appendage	*Alveophoma*	**G5**

Pseudopycnidia

126(119) Conidioma cupulate		127
Conidioma non-cupulate		131

---

*De Hoog (1985) puts the species with trilobate conidia in this genus and those with erect phragmoconidia in *Ochroconis*.

127(126) Conidia released in globular mass — *Bothrodiscus* — **G7**
          Non-agglomerated conidia — 128
128(127) Conidium with appendages — 129
          Conidium lacking appendages — 130
129(128) Apical appendage non-ramified, eccentric basal appendage sometimes present — *Heteropatella* — **G8**
          Apical appendage ramified — *Libartania* — **G24**
130(128) Conidium up to 4-septate, presence of a hypostroma — *Septopatella* — **G9**
          Conidium up to 7-septate, no hypostroma — *Oncosporella* — **G10**
131(126) Pseudopycnidium hysteroid, erumpent, elongated — *Moralesia* — **G11**
          No such character — 132
132(131) Conidioma flat, associated with blackening of tissue — 133
          Conidioma not flat — 138
133(132) Conidium unicellular — 134
          Conidium filiform, multicellular — 137
134(133) Conidium allantoid — *Leptothyrina* — **G12**
          Conidium non-allantoid — 135
135(134) Conidium oval-ellipsoid — *Leptostroma* — **G13**
          Conidium cylindrical — 136
136(135) Conidiogenous cell short — *Crandallia* — **G14**
          Conidiogenous cell filiform, often ramified — *Thoracella* — **G15**
137(133) Conidium up to 2-septate, thin at the tips — *Placothyrium* — (IV E14)
          Conidium up to 8-septate, truncated at the base — *Pleurothyrium* — (IV E17)
138(132) Conidioma hemispherical — *Leptochlamys* — **G16**
          Conidioma non-hemispherical — 139
139(138) Conidioma yellow-orange, hollowed out into locules resembling pycnidia — *Polystigmina* — **G17**
          Conidioma dark brown to black — 140
140(139) Conidium with appendages — *Furcaspora* — **G18**
          Conidia lacking appendages — 141
141(140) Conidium non-septate — 142
          Conidium multiseptate — 143
142(141) Conidium cylindrical — *Conostroma* — **G19**
          Conidium curved, filiform — *Tryblidiopycnis* — **G20**
143(141) Conidium up to 7-septate — *Stictosepta* — **G21**
          Conidium up to 3-septate — *Septocyta* — **G22**
144(118) Conidioma of sporodochium type — 145
          Conidioma of acervulus or coremium type — 151

Sporodochia _____

145(144) Sporodochium non-stromatic, atypical, conidia multiseptate, filiform — *Mycocentrospora* — **H5**

	Typical sporodochium more or less		
	stromatic		146
146(145)	Conidia elongated to filiform hyaline		147
	Conidia melanized		149
147(146)	Conidia unicellular	*Microdochium*	**H1**
	Conidia multicellular		148
148(147)	Conidia released in a gelatinous, yellow mass surrounded by the black bristles of the conidioma	*Wiesneriomyces*	**H4**
	No such characters	*Linodochium*	**H3**
149(146)	Conidia muriform	*Thyrostromella*	**H6**
	Conidia non-muriform		150
150(149)	Conidium globular, unicellular	*Hadrotrichum*	**H7**
	Conidium 0- to 3-septate, largely fusiform	*Fusicladium*	**H2**
151(144)	Conidioma acervular		152
	Coremium-type conidioma		159

Acervuli

152(151)	Acervulus subcuticular (cf. also *Pestalozziella* G4)		153
	Acervulus intra- or subepidermal		154
153(152)	Conidia similar to cat's paw print	*Vestigium*	**H9**
	Conidia unicellular, fusiform	*Fusamen*	**H10**
154(152)	Conidium straight, cylindrical or variously curved		155
	Conidium non-cylindrical		157
155(154)	Conidia non-septate, sometimes mixed with uniseptate macroconidia (microconidial form of *Phloeosporella* cf. infra)	*Microgloeum*	**H12**
	Conidia septate		156
156(155)	Conidia hyaline, 1- to 2-septate	*Diplosporonema*	**H8**
	Conidia hyaline to light brown, 0- to multiseptate	*Colletogloeum*	(V D4)
157(154)	Conidia guttulate, fusiform	*Rhabdogloeopsis*	**H11**
	Conidia filiform		158
158(157)	Conidia very curved, non-septate	*Libertella*	**H14**
	Conidia hardly or not curved, 1- to multiseptate	*Phloeosporella*	**H13**

Coremia

159(151)	Conidia rhomboidal	*Rhombostilbella*	**I 7**
	Conidia non-biconic		160
160(159)	Coremium poorly developed, small tuft of conidiophores on a stroma, conidia multicellular	*Pseudocercospora*	**I 10**
	Coremium well-developed		161
161(160)	Fertile zone on the whole of the coremium	*Isaria*	**I 1**

	Fertile zone on only part of the coremium		162
162(161)	Fertile zone on the upper part of the coremium		163
	Fertile zone limited to the middle or the capitulum		164
163(162)	Conidium unicellular, sometimes bicellular	*Phaeoisaria*	I 4
	Conidium multicellular	*Arthrosporium*	I 8
164(162)	Fertile zone on the middle part, sterile capitulum	*Chryseidea*	I 6
	Fertile zone limited to the capitulum		165
165(164)	Coremium divided at tip into several fertile branches	*Harpographium*	I 5
	No such character		166
166(165)	Conidium unicellular		167
	Conidium multicellular		168
167(166)	Conidiogenous cell monoblastic	*Graphium*	I 2
	Conidiogenous cell polyblastic	*Pachnocybe*	I 3
168(166)	Conidium curved into a hook		169
	Conidium straight	*Phaeoisariopsis*	I 9
169(168)	Conidia uniformly coloured with dark septal bands	*Trochophora*	I 11
	Conidia discoloured	*Curvulariopsis*	I 12

Table VI.A. Hyphales, Sympodulosporae

Fig.	Genus	No. spp.	Geographical distribution	Mode of life, substrates and impact	Teleomorphs	Additional references	Mol. biol.
1	*Helicomyces*	11	Cosmopolitan	Saprophyte on plants	*Ophionectria* *Tubeufia*	Mycol. 77: 606 (1985)	
2	*Helicosporium*	16	Cosmopolitan	Saprophyte on plants	*Ophionectria* *Tubeufia*	Mycol. 81: 356 (1989)	
3	*Helicoma*	33	Cosmopolitan	Saprophyte on plants	*Thaxteriella* *Tubeufia*	Mycol. 78: 744 (1986)	
4	*Helicomina*	7	Mostly tropical	Parasite on plants			
5	*Helicoon*	9	Cosmopolitan	Aero-aquatic and telluric fungus and on rotting plants		TBMS 87: 115 (1986); Myc. Res. 98:74 (1994)	
6	*Helicodendron*	19	Cosmopolitan, mostly temperate	Saprophyte on plants	*Capronia* *Hymenoscyphus* *Lambertella,* *Mollisia,* *Tyrannosorus*	TBMS 84: 423 (1985); TBMS 88: 63 (1987)	X
7	*Hiospira*	1	Tropical	Foliicolous	*Brooksia*		
8	*Helicorhoidion*	3	Cosmopolitan	Saprophyte on plants			
9	*Beltrania*	12	Cosmopolitan	Saprophyte on litter			
10	*Beltraniopsis*	3	Tropical	Saprophyte, foliicolous			
11	*Beltraniella*	8	Cosmopolitan, mostly tropical	Saprophyte on litter	*Leiosphaerella*		
12	*Ellisiopsis*	3	Cosmopolitan, mostly tropical	Saprophyte on plants			
13	*Pseudobeltrania*	4	Tropical	Saprophyte or leaf parasite			
14	*Hemibeltrania*	5	Cosmopolitan	Saprophyte or parasite on leaves and various plant organs			
15	*Rhombostilbella*	2	Brazil, Java	Parasite on ascomata of Capnodiaceae		Myc. Pap. 90 (1963)	

Hyphales, Sympodulospora

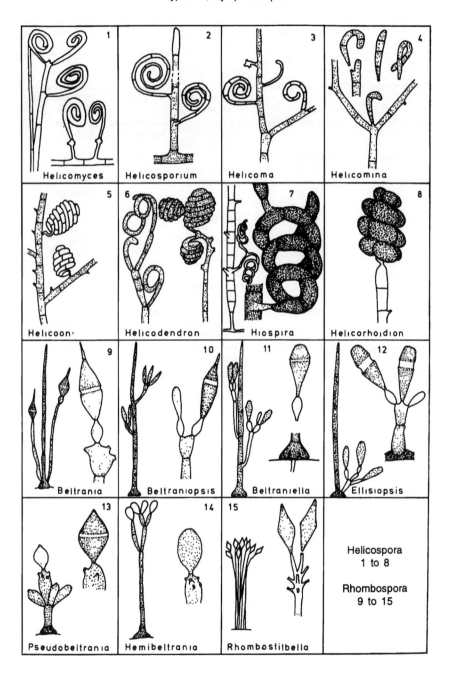

1 Helicomyces	2 Helicosporium	3 Helicoma	4 Helicomina
5 Helicoon·	6 Helicodendron	7 Hiospira	8 Helicorhoidion
9 Beltrania	10 Beltraniopsis	11 Beltraniella	12 Ellisiopsis
13 Pseudobeltrania	14 Hemibeltrania	15 Rhombostilbella	Helicospora 1 to 8 Rhombospora 9 to 15

Table VI.B. Nematophagous Hyphales, Sympodulosporae

Fig.	Genus	No. spp.	Geographical distribution	Mode of life, substrates and impact	Teleomorphs	Additional references	Mol. biol.
1	*Candelabrella*	6	Cosmopolitan	Telluric, nematophagous			
2	*Arthrobotrys*	28	Cosmopolitan	Telluric, nematophagous and on plants	*Orbilia*	Stud. Mycol. 26 (1985); 39 (1996)	
3	*Dactylaria*	40	Cosmopolitan	Parasite or saprophyte on plants, on humans	*Acrospermum, Hypomyces*	Stud. Mycol. 26 (1985)	
4	*Dactylella*	38	Temperate North	Telluric nematophagous, mycoparasite	*Orbilia*	Mycosyst. 7: 111 (1994) Stud. Mycol. 39 (1996)	
5	*Monacrosporium*	43	Cosmopolitan	Nematophagous	*Orbilia*	Myc. Res. 98: 862 (1994); Stud. Mycol. 39 (1996)	
6	*Geniculifera*	7	Cosmopolitan	Telluric, nematophagous and on plants		Study. Mycol. 26 (1985)	
7	*Nematoctonus*	16	Cosmopolitan	Telluric, nematophagous and on plants	*Hohenbuehelia* (B)	Mycol. 25: 321 (1986)	
8	*Duddingtonia*	1	Great Britain	Nematophagous and on litter		Stud. Mycol. 26 (1985)	
9	*Tridentaria*	2	Temperate	Nematophagous and on wood			

Nematophagous Hyphales, Sympodulosporae

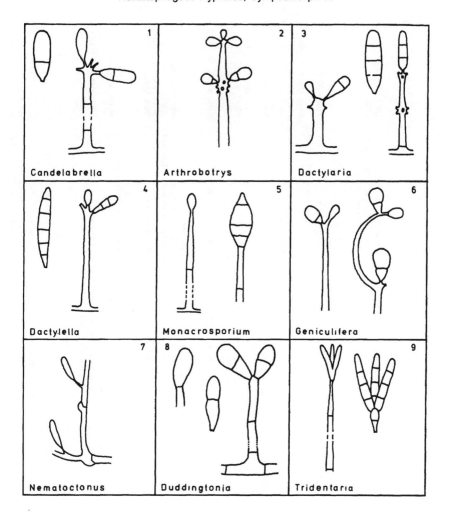

Candelabrella — Arthrobotrys — Dactylaria — Dactylella — Monacrosporium — Geniculifera — Nematoctonus — Duddingtonia — Tridentaria

Table VI.C. Amerosporous Hyphales Sympodulosporae

Fig	Genus	No. spp.	Geographical distribution	Mode of life. substrates and impact	Teleomorphs	Additional references	Mol. biol.
1	Lomentospora	1	Belgium	Tellunc and saprophyte			X
2	Geniculodendron	1	Canada, Gr. Britain	Parasite on seeds	Caloscypha		X
3	Sporothrix	39	Cosmopolitan	Saprophyte or parasite on plants, animals and humans. cases of hypovirulence	Calocera (B), Calosphaena. Cerinomyces (B), Dacrymyces (B), Ditiola (B), Dolichoascus. Fragosphaeria, Ophiostoma. Phaeosporis, Pseudeurotium. Stephanoascus	UP (1981): APS (1993)	X
4	Nodulisporium	28	Cosmopolitan	Saprophyte on wood and plant debris	Biscogniauxia. Hypoxylon, and other Xylariaceae + Graphostroma	TBMS 85: 391 (1985)	
5	Calcarisporella	1	Gr. Britain	Soil of slag heap		Stud. Mycol. 7 (1974)	
6	Calcarisporium	1	Europe	Mycoparasite		Stud. Mycol. 7 (1974)	
7	Costantinella	2	N. Amer.. Europe	Tellunc and on plants			X
8	Basifimbia	3	Eurasia	Coprophile	Morchella		
9	Beauvena	7	Cosmopolitan	Tellunc, entomopathogen and parasite on humans, used as biopesticide	Cordyceps	Stud. Mycol. 1 (1972)	X
10	Rhinotrichella	1	Temperate North	Mycoparasite		Stud Mycol. 15 (1977)	
11	Raffaelea	11	Cosmopolitan	Ambrosia fungus associated with *Platypodidae*		ASNT 50: 185 (1998)	
12	Leptodontidium	7	Temperate North, India	Lignicolous. rarely tellunc		Stud. Mycol. 15 (1977)	X
13	Rhinocladiella	7	Cosmopolitan	Generally lignicolous. sometimes parasite on humans	Caprona, Thecotheus	Stud. Mycol. 15 (1977)	X
14	Virgaria	1	Cosmopolitan	Corticolous. lignicolous	Hypoxylon		
15	Acrostaphylus	8	Ubiquitous	Saprophyte on plants	Entoleuca, Hypoxylon,		
16	Geniculosporium	12	Cosmopolitan	Saprophyte on plants. tellunc	Leprieuria. Nemania. Rosellinia	Myc. Res. 93: 121 (1989)	
17	Tritirachium	2	Cosmopolitan	Saprophyte or parasite on plants		Stud. Mycol. 1 (1972)	
18	Acrodontium	9	Cosmopolitan	Tellunc saprophyte or parasite on plants. animals and humans	Ascocorticium	Stud. Mycol. 1 (1972)	
19	Virganella	6	Ubiquitous	Tellunc, saprophyte on dead wood	Gibbera, Hypoxylon		
20	Acladium	23	Ubiquitous	Saprophyte on wood	Arachniotus, Botryobasidium (B)		
21	Idriella	16	Cosmopolitan	Tellunc parasite or saprophyte on litter and plants. endophyte	Lambertellina	Mycot. 8: 402 (1979)	
22	Leptographium	28	Cosmopolitan	Tellunc, pathogenic or saprophyte on woody plants	Ophiostoma	UP (1981): APS (1993)	X
23	Verticicladium	1	Ubiquitous	Saprophyte on needles of *Pinus* spp.	Desmazierella		
24	Pucola	1	Italy	Tellunc. saprophyte			
25	Hansfordia	9	Ubiquitous	Saprophyte on litter. mycoparasite	Geopyxis		
26	Dicyma	12	Ubiquitous	Tellunc saprophyte on litter	Ascotricha		
27	Conoplea	7	Ubiquitous	Parasite or saprophyte on woody plants	Korfiella, Plectania, Urnula		
28	Paraphaeosaria	1	USA	Parasite on *Cronartium*			
29	Selenosporella	± 11	Ubiquitous	Saprophyte on plants. mycoparasite	Diatrype Cookeina. Nanoscypha.	Mycol 25: 165 (1986).	
30	Molliardiomyces	10	Central and N Amer. Europe	Saprophyte on plants. rotting wood	Phillipsia. Pithya. Sarcoscypha	Mycot. 38: 417 (1990)	

Amerosporous Hyphales, Sympodulosporae

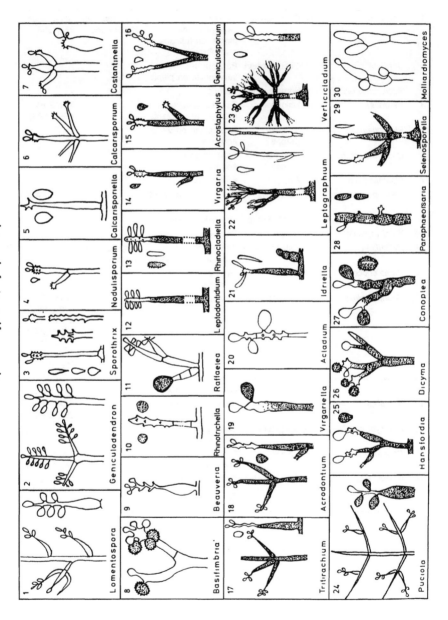

Table VI.D. Phragmosporous Hyphales, Sympodulosporae

Fig.	Genus	No. spp.	Geographical distribution	Mode of life, substrates and impact	Teleomorphs	Additional references	Mol. biol.
1	*Dactylaria*	41	Cosmopolitan	Parasite or saprophyte on plants, on humans, endophyte	*Acrospermum, Hypomyces*	Stud. Mycol. 26 (1985)	
2	*Cordana*	8	Cosmopolitan	Parasite or saprophyte on plants			
3	*Scolecobasidiella*	1	Somalia	Telluric			
4	*Veronaea*	16	Ubiquitous	Telluric, parasite or saprophyte on plants		TBMS 81: 485 (1983)	
5	*Pleurothecium*	1	N. Amer., Europe	Saprophyte on wood and bark			
6	*Cacumisporium*	2	N. Amer., Europe	Saprophyte on wood and bark			
7	*Blastophorum*	2	Asia	Foliicolous			
8	*Pleurophragmium*	15	Ubiquitous	Saprophyte on plants		Stud. Mycol. 26 (1985)	
9	*Dactylosporium*	2	Cuba, Europe	On rotting wood			
10	*Brachysporium*	9	Ubiquitous	Saprophyte on wood			
11	*Pyricularia*	9	Cosmopolitan	Pathogenic or saprophyte on plants	*Magnaporthe, Massarina*		X
12	*Paratrichoconis*	1	China	Hyperparasite of *Asterina*			
13	*Sporidesmiella*	18	Cosmopolitan	Saprophyte on plants		TBMS 79: 479 (1982); Mycot. 18: 243 (1983); Myc. Res. 102: 548 (1998)	
14	*Scolecobasidium*	25	Cosmopolitan	Telluric, on litter, plants, algae, fungi		Stud. Mycol. 26 (1985)	
15	*Subulispora*	17	Ubiquitous	Saprophyte on litter		Stud. Mycol. 26 (1985)	
16	*Parasympodiella*	3	Ubiquitous	Saprophyte on litter and branches			
17	*Camposporium*	15	Ubiquitous	Saprophyte on litter and wood		T.M.S. Jap. 34: 71 (1993)	
18	*Pleiochaeta*	3	Cosmopolitan	Pathogen of leaves and roots of *Lupinus*, or saprophyte on litter		Mycot. 57: 457 (1996)	

Phragmosporous Hyphales, Sympodulosporae

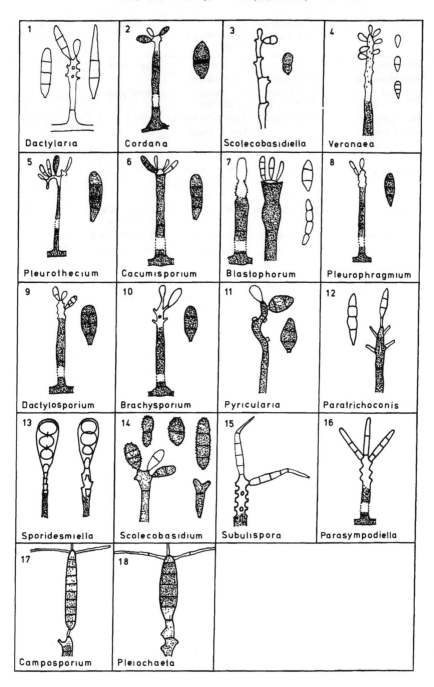

Table. VI.E. Phragmosporous Hyphales, Sympodulosporae

Fig.	Genus	No. ssp.	Geographical distribution	Mode of life, substrates and impact	Teleomorphs	Additional references	Mol. biol.
1	*Arnoldiomyces*	2	USA	Mycoparasite	*Hypomyces*		
2	*Pseudohansfordia*	9	Ubiquitous	Mycoparasite			
3	*Dactylium*	15	Ubiquitous temperate	Lignicolous, fongicolous	*Hypomyces*		
4	*Pseudobotrytis*	1	Ubiquitous	Telluric, on plants			
5	*Parapyricularia*	1	Asia	Leaf of *Musa*			
6	*Sterigmatobotrys*	2	Ubiquitous	Saprophyte on wood			
7	*Spondylocladiopsis*	1	Europe	Saprophyte on cupules of *Fagus*			
8	*Sirosporium*	14	Ubiquitous	Leaf parasite			
9	*Cercospora*	± 1300	Cosmopolitan	Phytopathogen	*Mycosphaerella, Sphaerulina*		X
10	*Cercosporidium*	± 27	Cosmopolitan	Phytopathogen	*Mycosphaerella,*		
11	*Passalora*	15	Ubiquitous	Leaf parasite	*Mycosphaerella*	Mycot. 51: 509 (1994)	
12	*Pseudocercosporella*	10	Ubiquitous	Phytopathogen (eyespot of cereals)	*Mycosphaerella, Tapesia*		X

Phragmosporous Hyphales, Sympodulosporae

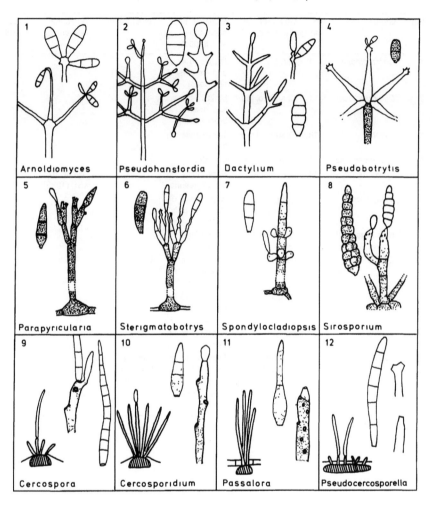

Table VI.F. Hyphales, Sympodulosporae

Fig.	Genus	No. spp.	Geographical distribution	Mode of life, substrates and impact	Teleomorphs	Additional references	Mol. biol.
1	*Ceratocladium*	1	Europe	On wood and bark (mostly of *Fagus*)			
2	*Circinotrichum*	11	Cosmopolitan	Mostly litter and dead branches			
3	*Gyrotrix*	18	Cosmopolitan	Saprophyte on plants			
4	*Kramasamuha*	1	S. Amer., India	On litter			
5	*Polyschema*	14	Ubiquitous	Saprophyte on plants, telluric		Mycot. 57: 451 (1996)	
6	*Dwayabeeja*	1	India	Saprophyte on rachis of palm trees			
7	*Kumanasamuha*	1	Africa, India	On rotting wood			
8	*Zygophiala*	1	Ubiquitous	Agent of fly speck disease of apples	*Schizothyrium*		
9	*Zygosporium*	9	Ubiquitous	Saprophyte or parasite on plants			
10	*Balanium*	1	Europe	Saprophyte on wood			
11	*Xylocladium*	6	Ubiquitous mostly America	Saprophyte or weak parasite, on wood	*Camillea, Jongiella, Poronia*		
12	*Diplocladiella*	5	Ubiquitous mostly tropical	Saprophyte on plants		Mycot. 46: 235 (1993)	

Hyphales, Sympodulosporae

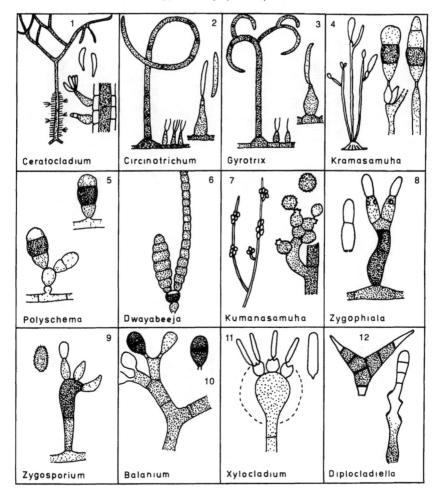

Table VI.G. Pycnidial and pseudopycnidial Sympodulosporae

Fig.	Genus	No. spp.	Geographical distribution	Mode of life, substrates and impact	Teleomorphs	Additional references	Mol. biol.
1	Eriospora	5	S. Amer., Europe	On dead leaves of Carex, Juncus, Typha Berberis....	Stictis		
2	Cornutispora	5	N. Amer., Australia, Europe	Hyperparasite on thallus of Parmelia (lichen), apothecium of Therrya (on branch of Pinus) and other Rhytismatales			
3	Septoria	1000	Cosmopolitan	Many species, often pathogenic, maculicolous, chiefly on leaves, of many plants	Mycosphaerella, Phaeosphaeria, Sphaerulina	Mik. Fit. 26: 425 (1992); Mycol. 83: 611 (1991); O.U.P. (1965, 1967); Syd. 44: 13 (1992); TBMS 83: 383 (1984); TTB 1987	X
4	Pestalozziella	4	Eurasia, USA	On living or dead leaves of various plants			
5	Alveophoma	1	Spain	On dead twigs of Fontanesia			
6	Dendroseptoria	2	Spain, Portugal	Maculicolous on culm of Poaceae			
7	Bothrodiscus	3	N. Amer., Asia	On branches of firs (Abies spp.)	Ascocalyx	N. Hedw. 34: 82 (1981)	
8	Heteropatella	5	Europe, USA	On dead stems of many leafy plants, 1 sp. parasite on Antirrhinum	Heterosphaeria		
9	Septopatella	1	N. Amer., Europe	On needles of pine (Pinus spp.)			
10	Oncosporella	1	Finland	On dead wood of Populus			
11	Moralesia	1	Canary Islands	On dry culms of Arundo donax			
12	Leptothyrina	1	Switzerland	Agent of leaf necrosis on Rubus			
13	Leptostroma	200	Ubiquitous	On culms or leaves of various plants, 9 spp. parasites on Pinus, endophyte	Hypoderma, Lophodermium	Myc. Pap. 147 (1981); Mycol. 83: 89 (1991)	
14	Crandallia	1	USA	On living stems and dead leaves of Juncus	Duplicaria		
15	Thoracella	1	Europe	Agent of tar spot on leaves of Ledum			
16	Leptochlamys	1	Europe	On mosses			
17	Polystigmina	2	Eurasia	Agent of leaf lesions on Prunus spp.	Polystigma		
18	Furcaspora	1	USA	On dead needles of Abies and Pinus			
19	Conostroma	2	Eurasia, USA	On branches of Quercus	Colpoma		
20	Tryblidiopycnis	1	N. Amer., Eurasia	Endophyte, and on dead stems, branches and bark of Picea spp., sometimes on Pinus and Larix	Tryblidiopsis	CJB 72: 549 (1994)	
21	Stictosepta	1	Czechoslovakia	On dead branches of Fraxinus			
22	Septocyta	2	Europe, India	On stems of Rubus spp. and on trunk of Corchorus	Arthonia	Myc. Res. 99: 166 (1995)	
23	Pocillopycnis	1	Sweden	On twigs of Picea abies		Mycot. 10: 288 (1980)	
24	Libartania	3	Australia, Canada, Italy	Stems of Laserpitium and culms (Phragmites, Themeda)	Phragmiticola		

Pycnidial and pseudopycnidial Sympodulosporae

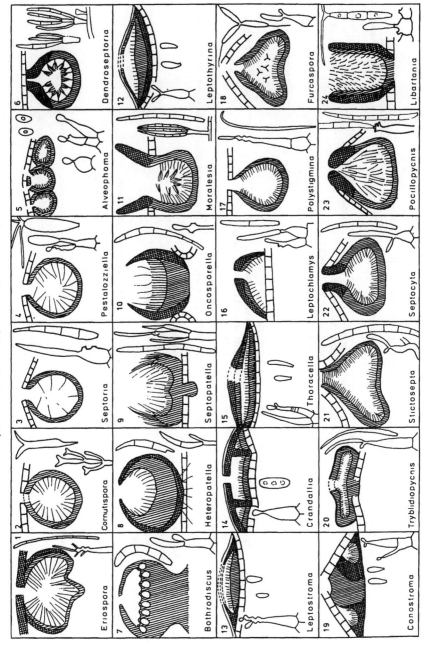

Table VI.H. Sympodulosporae with sporodochia and acervuli

Fig.	Genus	No. spp.	Geographical distribution	Mode of life, substrates and impact	Teleomorphs	Additional references	Mol. biol.
1	*Microdochium*	12	Cosmopolitan	Pathogen or saprophyte on various plant organs, telluric, and on oospores of *Phytophthora*, 2 spp. on insects and faeces	*Monographella*	N. Hedw. 27: 215 (1976); Syd. 34: 30 (1981)	X
2	*Fusicladium*	40	Cosmopolitan	Generally a parasite on various organs of many plants	*Acantharia, Apiosporina, Microcyclus, Venturia*		
3	*Linodochium*	1	Europe, USA	On needles of *Pinus* spp. and leaves of *Quercus*		CJB 57: 370 (1979)	
4	*Wiesneriomyces*	3	Ubiquitous, mostly tropical	Saprophyte on rotted leaves and stems		TBMS 90: 619 (1988)	
5	*Mycocentrospora*	14	Cosmopolitan	Parasite or saprophyte on various living organs (leaf necrosis) or dead organs of many plants, telluric, aquatic		Mycot. 54: 37 (1995)	
6	*Thyrostromella*	1	Australia, Europe	On leaves and culms of *Ammophila*			
7	*Hadrotrichum*	5	Ubiquitous, mostly temperate zone	Telluric, parasite or saprophyte (follicolous, caulicolous, lignicolous)	*Hypoxylon, Kretzschmaria*	Mycot. 18: 91 (1983)	
8	*Diplosporonema*	1	Ubiquitous	On leaves and stems of Caryophyllaceae			
9	*Vestigium*	1	Canada, USA	On twigs of *Thuja plicata*			
10	*Fusamen*	5	Europe	On leaves and catkins of *Salix* spp.			
11	*Rhabdogloeopsis*	1	USA	On needles of firs (*Abies* spp.)			
12	*Microgloeum*	1	Ubiquitous	Pathogen on branches and leaves of *Prunus* spp.	*Blumeriella*		
13	*Phloeosporella*	5	N. and S. Amer., Eurasia	Pathogen on living leaves of various woody plants including *Prunus* spp. (agent of anthracnose)	*Blumeriella*		
14	*Libertella*	20	Temperate zone	Saprophyte on various organs of many woody plants	*Barrmaelia, Creosphaeria, Cryptosphaeria, Diatrypella, Lopadostoma*		

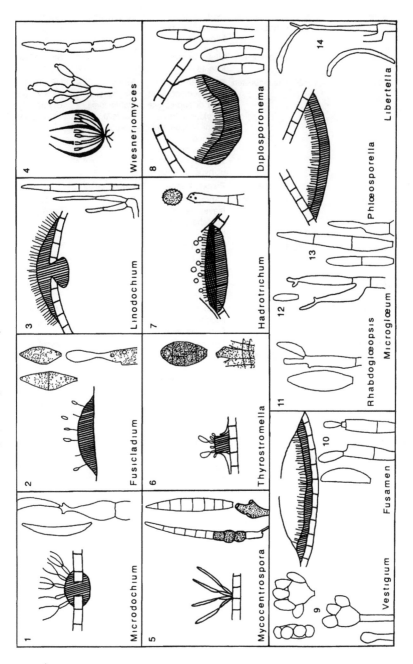

Sympodulosporae with sporodochia and acervuli

1 Microdochium 2 Fusicladium 3 Linodochium 4 Wiesneriomyces

5 Mycocentrospora 6 Thyrostromella 7 Hadrotrichum 8 Diplosporonema

9 Vestigium 10 Fusamen 11 Rhabdogloeopsis 12 Microgloeum 13 Phloeosporella 14 Libertella

Table VI.I. Sympodulosporae with coremia

Fig.	Genus	No. spp.	Geographical distribution	Mode of life, substrates and impact	Teleomorphs	Additional references	Mol. biol.
1	*Isaria*	2	Ubiquitous	Litter, faeces, insects		Stud. Mycol. 1 (1972)	
2	*Graphium*	30	Cosmopolitan	Pathogen (tracheomycosis of elm) or saprophyte on plant debris, agent of blue stain of wood, dissemination by xylophages, cases of hypovirulence	*Kernia, Microascus, Ophiostoma, Petriella, Pseudallescheria*	UP (1981); APS (1993); Crypto. Myc. 14: 219 (1993)	X
3	*Pachnocybe*	1	Europe	Planks, beams	*Ustomycetes* (B)		
4	*Phaeoisaria*	10	Cosmopolitan	Saprophyte on leaves, wood, branches, stems, bark	*Scoptria*	Pers. 8: 407 (1976)	
5	*Harpographium*	10	Ubiquitous	Saprophyte on wood, dead branches			
6	*Chryseidea*	1	Africa	Saprophyte on dead leaves			
7	*Rhombostilbella*	2	Brazil, Java	Parasite on ascomata of Capnodiaceae		Myc. Pap. 90 (1963)	
8	*Arthrosporium*	2	N. Amer., Southeast Asia, Europe	Saprophyte on wood and fruits		Mycol. 64: 1175 (1972)	
9	*Phaeoisariopsis*	14	Ubiquitous, mostly, tropical	Mostly leaf parasite	*Mycosphaerella*	Myc. Res. 95:1005 (1991)	X
10	*Pseudocercospora*	± 250	Ubiquitous, mostly tropical	Leaf parasite	*Mycosphaerella*	Mycot. 82: 313 (1990)	X
11	*Trochophora*	1	Asia	Living leaves of *Daphniphyllum*			
12	*Curvulanopsis*	1	Central Amer.	Leaves of *Rubus*			

Sympodulosporae with coremia

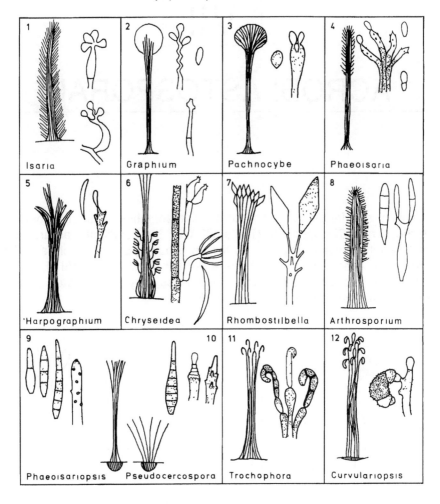

1 Isaria	2 Graphium	3 Pachnocybe	4 Phaeoisaria
5 'Harpographium	6 Chryseidea	7 Rhombostilbella	8 Arthrosporium
9 Phaeoisariopsis	10 Pseudocercospora	11 Trochophora	12 Curvulariopsis

VII

ACROBLASTOSPORAE*

A typical and frequently encountered representative of Acroblastosporae is the genus *Cladosporium* (Table VII.A 2): the conidiophores are more or less differentiated hyphae. At the tip a bud appears, which is at first minuscule, then enlarges until it is a cellular element (Fig. VII.1). This in turn buds into other cells, and the process may be repeated a great number of times (Hashmi *et al.*, 1973).

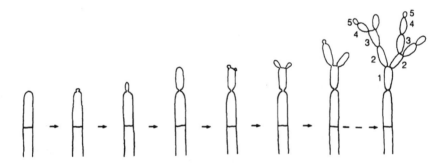

Fig. VII.1. Conidiogenesis by successive budding in an Acroblastosporae (*Cladosporium*).

Each of the cells thus produced is an element of asexual reproduction, a conidium, capable of germinating and giving rise to a new colony of *Cladosporium*. Finally, at the tip of the conidiophore, there is a group of chains of conidia.

This basic outline calls for certain remarks and presents some variations:

— At points of budding, the link between conidia is fragile (it is difficult to observe an intact, entire sporiferous apparatus);

---

*The term **Acroblastosporae** was suggested to us by Professor Hennebert to designate this group of fungi, which we had earlier labelled Blastosporae *sensu stricto*.

— The detached conidia have scars at the points at which they were attached to the conidiophores or to lower or higher conidia. Certain genera have disjunctors between conidia that, at maturity, enable the release of the conidia and leave a fringe at the level of the hilum and the conidial scar (e.g., *Junctospora*, Table VII.A 18).

— A conidium may bud at two or several points towards its tip, and the conidial chain is thus ramified. Also, if one observes a detached conidium with two scars at the tip, that definitely indicates that the fungus is a Blastosporae s.l., sporulating according to the model described above (fundamental difference with the Phialidae, in which the chains can only be simple).

— The genus *Cladosporium* comprises fungi with melanized wall. The presence of melanin renders the scars more visible. In the genus *Hyalodendron* (Table VII.A 3), colourless homologue of *Cladosporium*, they are less marked. Inversely, in certain black fungi, conidiophore and conidium may have an annular thickening around the scar: even though their mode of conidial production is very similar to that described here, they are classed in a special section, Porosporae (see chap. IX).

— In *Cladosporium*, simple conidia (amerospores) are found, or they have one or several septa (didymospores, phragmospores) (De Vries, 1952; Ellis, 1971a). In other genera (*Polyscytalum*, Table VII.B 4; *Septonema*, Table VII.B 3), the presence of a determined number of septa in mature conidia is the rule.

— In *Periconia* (Table VII.A 5), we can see a remarkable peculiarity: the conidia are produced in basifugal chains, but the maturation is basipetal, beginning at the youngest conidium and proceeding backwards in the direction of the conidiophore.

— The genus *Bispora* (Table VII.A 19), for example, has simple chains of conidia with a wide scar, suggesting a succession of aleuriospores.

— No Acroblastosporae are found in the Conidiomales that have pycnidia and acervuli. In fact, the difficulties of observing conidia in chains do not allow us always to determine the mode of conidiogenesis as arthric or blastic (Sutton, 1973a). We have followed the system of Sutton, who, in his fundamental work on the Coelomycetes (Sutton, 1980), places almost all these organisms among the Arthrosporae.

On the other hand, some genera of Conidiomales with coremia or sporodochia are represented, as well as the pulvinate intermediate forms (Table I.F f2).

Key to identification of Acroblastosporae

1	Hyphales	2
	Conidiomales	36

Table VII.A. Acroblastosporae

Fig.	Genus	No. spp.	Geographical distribution	Mode of life, substrates and impact	Teleomorphs	Additional references	Mol. biol.
1	*Alysidium*	3	Ubiquitous, Northern Hemisphere	Saprophyte on dead wood			
2	*Cladosporium*	± 50	Cosmopolitan	Saprophyte on many substrates, phytopathogen (cladosporiosis, scab), hyperparasite of Uredinales, toxinogenic	*Apiosporina, Capronia, Lasiobotrys, Mycosphaerella, Venturia, Xenomeris*	Myc. Pap. 172 (1998)	X
3	*Hyalodendron*	1	Ubiquitous	Lignicolous		Stud. Mycol. 19 (1979)	
4	*Monilia*	10	Ubiquitous	Agent of brown rot of flowers, branches and fruits, or of leaf spots on Rosaceae	*Monilinia*		X
5	*Periconia*	30	Cosmopolitan	Saprophyte on plant debris, telluric, pathogen on *Sorghum*	*Didymosphaeria*		
6	*Fulvia*	2	Cosmopolitan	Phytopathogen			X
7	*Ramularia*	300	Ubiquitous, mostly temperate	Agent of leaf spots on many plants	*Leptotrochila, Melanodothis, Mycosphaerella*	N. Hedw. 58:191 (1994)	
8	*Phaeoramularia*	26	Ubiquitous	Leaf parasite	*Mycosphaerella*		
9	*Stenella*	14	Cosmopolitan	Leaf parasite	*Mycosphaerella*	TBMS 81: 485 (1983)	
10	*Trichosporonoides*	5	N. Amer., Southeast Asia	Saprophyte		Mycopath. 108:25 (1989)	
11	*Xylohypha*	7	Ubiquitous, temperate	Saprophyte on plants, 1 sp. pathogen on humans			
12	*Dimorphospora*	1	Ubiquitous	Saprophyte on submerged leaves			
13	*Sphaerosporium*	2	Europe, USA	1 lignicolous, 1 keratinophilous			
14	*Mycosylva*	3	Eurasia, temperate and boreal zones	Coprophile			
15	*Pycnostysanus*	2	N. and Central Amer., Europe	Telluric, 1 sp. pathogenic on leaves and buds of *Rhododendron*		Agronomie 1:87 (1981)	
16	*Haplographium*	15	Cosmopolitan	Saprophyte, epiphyte	*Hyaloscypha*	Mycot. 46: 11 (1993)	
17	*Setodochium*	1	S. Amer.	Foliicolous			
18	*Junctospora*	1	Europe	Saprophyte on needles of *Pinus*			
19	*Bispora*	6	Ubiquitous	Saprophyte on plants			
20	*Oedemium*	2	Ubiquitous	Saprophyte on dead branches	*Chaetosphaerella, Thaxteria*		

Acroblastosporae

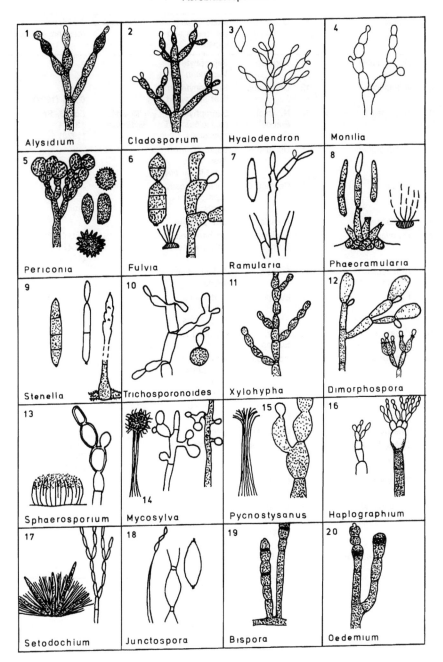

Table VII.B. Acroblastosporae

Fig.	Genus	No. spp.	Geographical distribution	Mode of life, substrates and impact	Teleomorphs	Additional references	Mol. biol.
1	Scopulariella	1	Denmark	Saprophyte on Vaccinium			
2	Anungitea	13	Ubiquitous	Saprophyte on litter and wood		Mycot. 65:93 (1997)	
3	Septonema	10	N. Amer., Europe	Saprophyte, corticolous and lignicolous	Ciliolarina, Mytilinidion	Fol. G. Pr. 13:421 (1978)	
4	Polyscytalum	8	Ubiquitous	Saprophyte on litter, 1 sp. parasite on potato			
5	Torula	18	Cosmopolitan	Telluric, mostly saprophyte on plants			
6	Haplobasidion	5	Eurasia	Leaf parasite			
7	Hormiactella	2	Antilles, Europe	On bark and rotting wood		Fol. G. Pr. 13:421 (1978)	
8	Hormiactis	5	Eurasia, USA	Telluric, plant debris			
9	Lylea	2	Europe	Corticolous, lignicolous, fongicolous		Fol. G. Pr. 13:421 (1978)	
10	Lobatopedis	3	Africa, Europe	On dead leaves			
11	Taeniolella	21	Ubiquitous	Mostly saprophyte on plants, lichenicolous			
12	Peyronelia	4	N. Amer., Europe	Corticolous and lignicolous	Glyphium		
13	Rhexoampullifera	1	Gr. Britain	Litter			
14	Sporophiala	3	India, USA	Lignicolous		TBMS 78:559 (1982)	
15	Taeniolina	1	Cuba, Europe	Foliicolous			
16	Septotrullula	1	Europe	Corticolous			
17	Pleurotheciopsis	3	S. Amer., Europe	Corticolous, lignicolous and on litter			
18	Parapleurotheciopsis	2	Gr. Britain, Japan	On litter			

Acroblastosporae

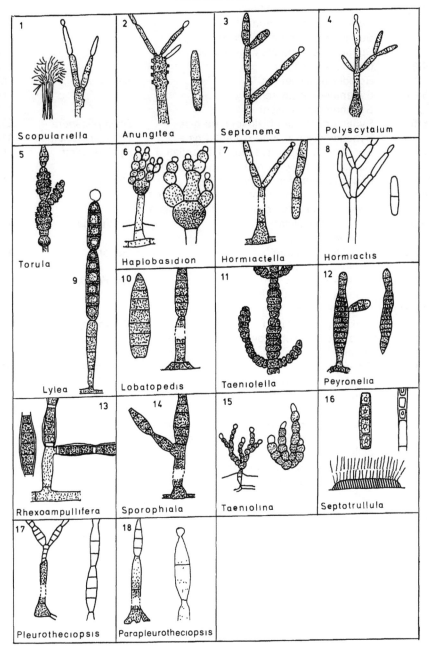

1 Scopulariella	2 Anungitea	3 Septonema	4 Polyscytalum
5 Torula	6 Haplobasidion	7 Hormiactella	8 Hormiactis
9 Lylea	10 Lobatopedis	11 Taeniolella	12 Peyronelia
13 Rhexoampullifera	14 Sporophiala	15 Taeniolina	16 Septotrullula
17 Pleurotheciopsis	18 Parapleurotheciopsis		

Hyphales

2(1)	Conidial chains ramified		3
	Conidial chains non-ramified		24
3(2)	Sporiferous apparatus hyaline		4
	Sporiferous apparatus melanized		8
4(3)	Conidia unicellular		5
	Conidia bicellular	*Ramularia*	**A7**
5(4)	Synanamorph present		6
	Synanamorph absent		7
6(5)	Synanamorph of phialosporous type	*Dimorphospora*	**A12**
	Synanamorph of arthrosporous type	*Trichosporonoides*	**A10**
7(5)	Sporiferous apparatus well-differentiated	*Hyalodendron*	**A3**
	Conidia seeming to arise from swelling and fragmentation of hypha	*Monilia*	**A4**
8(3)	Conidiophores poorly or not differentiated		9
	Conidiophores more or less differentiated		15
9(8)	Conidia unicellular	*Xylohypha*	**A11**
	Conidia septate		10
10(9)	Conidia uniseptate	*Bispora*	**A19**
	Conidia generally multiseptate		11
11(10)	Conidia ramified	*Taeniolina*	**B15**
	Conidia non-ramified		12
12(11)	Phragmoconidia with smooth outline, rhexolytic separation	*Rhexoampullifera*	**B13**
	Phragmo- or dictyoconidia with torulose outline		13
13(12)	Dictyoconidia	*Peyronelia*	**B12**
	Torulose phragmoconidia		14
14(13)	Homogenous chains of uniformly coloured phragmoconidia	*Taeniolella*	**B11**
	Chains of conidia 0- to multiseptate, often discoloured and often with conidiogenous cells having a dark base	*Torula*	**B5**
15(8)	Conidiophores semi-micronematous		16
	Conidiophores macronematous		17
16(15)	Compact conidiophores with chains of large cylindrical and distoseptate phragmoconidia	*Lylea*	**B9**
	Conidia bi- to multicellular, cylindrical	*Polyscytalum*	**B4**
17(15)	Conidia and conidiogenous cells arranged in terminal penicilli		18
	No such character		20
18(17)	Basipetal conidial maturation	*Periconia*	**A5**
	Acropetal maturation		19
19(18)	Conidia large, dry	*Haplobasidion*	**B6**
	Conidia small, mucous	*Haplographium*	**A16**
20(17)	Hilum and scar rather wide		21
	Hilum and scar rather narrow		22

21(20)	Conidia unicellular, very melanized at maturity	*Alysidium*	A1
	Conidia multicellular, fusoid	*Sporophiala*	B14
22(20)	Conidia of regular form and septation	*Septonema*	B3
	Conidia of irregular form and septation		23
23(22)	Lateral swellings of conidiophore	*Fulvia*	A6
	No such character	*Cladosporium*	A2
24(2)	Sporiferous apparatus hyaline	*Hormiactis*	B8
	Sporiferous apparatus melanized		25
25(24)	Conidia hyaline		26
	Conidia melanized		27
26(25)	Conidia unicellular with disjunctor	*Junctospora*	A18
	Conidia multicellular without disjunctor	*Pleurotheciopsis*	B17
27(25)	Conidia bicellular		28
	Conidia multicellular		31
28(27)	Chain of conidia on a 'grater' (:raduliform) head	*Anungitea*	B2
	No such character		29
29(28)	Conidia on rounded conidiogenous cells	*Oedemium*	A20
	No such character		30
30(29)	Conidia compact with distinctive septa	*Bispora*	A19
	Conidia narrow with slight melanization	*Hormiactella*	B7
31(27)	Conidiophore or conidiogenous cell rises from a massive, often lobed base		32
	No such character		33
32(31)	Conidia subhyaline	*Parapleurotheciopsis*	B18
	Conidia melanized	*Lobatopedis*	B10
33(31)	Conidia muriform	*Peyronelia*	B12
	Conidia phragmiate		34
34(33)	Conidia distoseptate	*Lylea*	B9
	Conidia euseptate		35
35(34)	Conidiophores formed from a substomatic stroma, smooth conidia	*Phaeoramularia*	A8
	Mycelium and conidiophores superficial, rough, conidia often echinulate	*Stenella*	A9

Conidiomales

36(1)	Coremia		37
	Sporodochia		39
37(36)	Conidia septate, cylindrical	*Scopulariella*	B1
	Conidia non-septate		38
38(37)	Capitulum spherical	*Mycosylva*	A14
	Capitulum fan-shaped	*Pycnostysanus*	A15
39(36)	Sporodochium radiate with melanized bristles	*Setodochium*	A17
	Sporodochium subcylindrical		40
40(39)	Ramified chains of large amerospores	*Sphaerosporium*	A13
	Simple chains of phragmoconidia	*Septotrullula*	B16

VIII

BOTRYOBLASTOSPORAE

The group Botryoblastosporae is close to the Sympodulosporae (chapter VI) and Acroblastosporae (chapter VII), but here the conidia are produced simultaneously in dense bunches, generally on an apical swelling of the conidiophore (ampulla).

— If the development is arrested at this stage, the result is a head with single conidia, as in *Oedocephalum* (Table VIII.3). The conidia may themselves bud into new propagules and form basifugal chains (*Gonatobotryum*, Table VIII.1). In this group they are almost exclusively amerospores, except in *Cephaliophora* (Table VIII.13) and *Engelhardtiella* (Table VIII.14).

— The sporogenous zones, instead of being more or less spherical like those above, may be elongated (*Chromelosporium*, Table VIII.11; *Myriodontium*, Table VIII.12).

— The conidiophore may proliferate, the first ampulla becomes intercalary, and new heads are produced (*Gonatobotrys*, Table VIII.2).

— The conidiophores may be ramified more or less regularly in arbuscules, each branch having a fertile head at its tip (*Botrytis*, Table VIII.10).

— Finally, the conidiophores may be grouped in fruit bodies (Table VIII.15 and 16).

The character that mainly defines this group is the simultaneous production of the primary conidia, the ultrastructure of which has been studied for certain genera in chapter III.

In practice, Barron (1968) differentiated the ampulla of Botryoblastosporae from certain swellings of Sympodulosporae:

— in the first case, the ampulla is produced before conidiogenesis;

— in the Sympodulosporae, the bulging is the result of successive conidiogenesis.

Blastobotrys is a particular case studied by de Hoog (1974) and Cole and Samson (1979). The conidiophore, which grows sympodially, produces ampullae, which are perhaps primary conidia, giving rise to secondary conidia more or less simultaneously. The secondary conidia remain attached to the ampullae and are released along with them.

The complex *Botrytis, Chromelosporium, Ostracoderma* has been discussed in detail by Hennebert (1973).

Key to identification of Botryoblastosporae

1	Hyphales		2
	Conidiomales		14
2(1)	Conidia unicellular		3
	Conidia multicellular	*Cephaliophora*	VIII 13
3(2)	Conidiophore simple, terminating in a single fertile ampulla		4
	Conidiophore with growths or branches		6
4(3)	Fungi always hyaline		5
	Fungi melanized with age	*Aureobasidium**	VIII 7A
5(4)	Fairly large fertile ampulla, more or less spherical, often separated from conidiophore by a septum; genus related to Ascomycetes	*Oedocephalum*	VIII 3
	Fertile swelling in the form of a club, not separated from conidiophore by a septum; genus related to Basidiomycetes	*Spiniger*	VIII 4
6(3)	Conidiophore with growths		7
	Conidiophore branched		10
7(6)	Percurrent proliferation of conidiophore		8
	Sympodial growths	*Blastobotrys*	VIII 6
8(7)	Conidia single	*Gonatobotrys*	VIII 2
	Conidia in basifugal chains		9
9(8)	Fungus hyaline	*Nematogonum*	VIII 5
	Fungus melanized	*Gonatobotryum*	VIII 1
10(6)	Branches by fours, at right angles, each carrying a fertile ampulla	*Sphondylocephalum*	VIII 8
	Branches of different types		11
11(10)	Fertile zones short, apical		12
	Fertile zones elongated		13
12(11)	Conidiophore dichotomic, with lateral, cruciate branches with fertile ampulla	*Botryosporium*	VIII 9
	Branches less regular, forming terminal bunches of fertile cells	*Botrytis*	VIII 10
13(11)	Conidial zones on poorly differentiated hyphae, very well-developed conidial denticles	*Myriodontium*	VIII 12

Continued on p. 152

*In yeast-like fungi, see also *Moniliella* (II A 12) and *Trichosporonoides* (VII A 10).

Table VIII. Botryoblastosporae

Fig.	Genus	No. spp.	Geographical distribution	Mode of life, substrates and impact	Teleomorphs	Additional references	Mol. biol.
1	Gonatobotryum	3	N. Amer., Europe	Saprophyte, myco- and phytoparasite		TBMS 77: 299 (1981)	
2	Gonatobotrys	2	Cosmopolitan	Myco- and phytoparasite	Melanospora	TBMS 77: 299 (1981)	
3	Oedocephalum	10	Cosmopolitan	Saprophyte and coprophile	Ascophanus, Cleistoiodophanus Iodophanus, Peziza, Pyronema	KNAWP 77: 383 (1974)	
4	Spiniger	16	Cosmopolitan	Saprophyte on wood, pathogen on trees	Heterobasidion (B) and other Hymenomycetes	KNAWP 77: 402 (1974)	X
5	Nematogonum	2	N. Amer., Eurasia	Mycoparasite		TBMS 77: 299 (1981)	
6	Blastobotrys	3	Europe	Saprophyte			
7a	Aureobasidium	7	Cosmopolitan	Saprophyte, phytoparasite, endophyte, pathogen on humans		Stud. Mycol. 15: 141 (1977)	X
7b	Kabatiella	15	Ubiquitous, temperate North	Saprophyte or parasite, mostly foliicolous	Discosphaerina, Sarcotrochila	Stud. Mycol. 15: 141 (1977)	
8	Sphondylocephalum	1	USA	Coprophile			
9	Botryosporium	4	Africa, N. Amer., Eurasia	Saprophyte, weak parasite			
10	Botrytis	50	Cosmopolitan	Saprophyte or pathogen on many economically important plants, toxinogen	Botryotinia		X
11	Chromelosporium	10	Cosmopolitan	Telluric, saprophyte on plant debris	Peziza		
12	Myriodontium	1	Europe, USA	Telluric, keratinophile			
13	Cephaliophora	6	Ubiquitous	Telluric, coprophile, predator of nematodes, rotifers and tardigrades			
14	Engelhardtiella	1	N. Amer.	Mycoparasite on Botryosphaeria			
15	Microstroma*	5	Ubiquitous	Leaf parasite on Quercus, Juglans, Albizzia, Castanopsis...	(B)		X
16	Ostracoderma	4	Ubiquitous	Telluric			

*TEM studies have shown the affinity of this genus with the Basidiomycetes (Mycol. 74: 285, 1982).

Botryoblastosporae

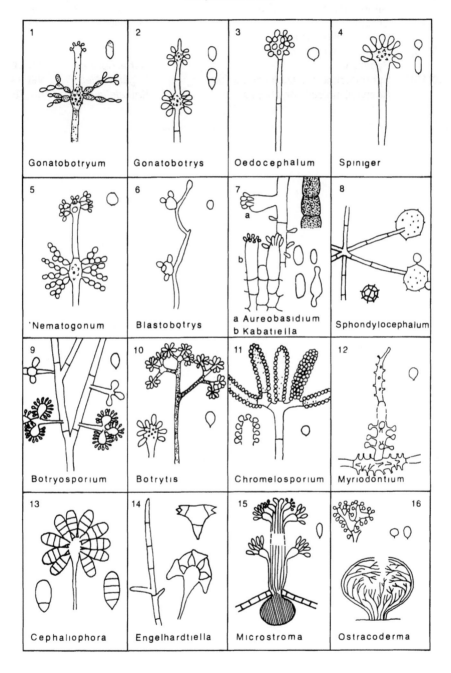

1 Gonatobotryum	2 Gonatobotrys	3 Oedocephalum	4 Spiniger
5 Nematogonum	6 Blastobotrys	7 a Aureobasidium b Kabatiella	8 Sphondylocephalum
9 Botryosporium	10 Botrytis	11 Chromelosporium	12 Myriodontium
13 Cephaliophora	14 Engelhardtiella	15 Microstroma	16 Ostracoderma

	Conidial zones at the tip of well-differentiated conidiophores	*Chromelosporium**	**VIII** 11
14(1)	Coremia	*Microstroma*	**VIII** 15
	Other conidiomata		15
15(14)	Sporodochia or acervuli		16
	Pycnidia	*Ostracoderma*	**VIII** 16
16(15)	Sporodochia, multicellular conidia	*Engelhardtiella*	**VIII** 14
	Acervuli, unicellular conidia	*Kabatiella*	**VIII** 7B

---

*Certain *Chromelosporium* have coremia.

IX

POROSPORAE

The subdivision Porosporae constitutes one of the original creations of modern taxonomy. Hughes (1953) defined section VI thus: 'Conidia ... developing from pores on conidiophores of determinate or indeterminate length; they are solitary or in whorls, and may occur in acropetal chains.'

As we have seen earlier in the introduction to the chapter on Blastosporae s.l., the characteristic pore of the group consists of a thin duct opening into the wall of the conidiogenous cell and surrounded by a melanized annular thickening. This structure of the conidial scar is generally found on the hilum. The Porosporae are all melanized fungi.

The conidia are called porospores or, according to the terminology proposed by Ellis (1971b) at Kananaskis, tretoconidia (from the Greek *treta:* canal).

In the classic conception of many authors, the pore or duct becomes visible by enzymatic dissolution in an adult wall, having acquired its definitive thickness and melanization. However, such a phenomenon is difficult to follow in the living fungus and it is not generally possible to determine whether the characteristic structure of the pore is formed before, during, or after conidiogenesis. Moreover, in this conception, the porospores should be enteroblastic, whereas studies to date under electron microscope show that they are holoblastic, as seen earlier (chapter III).

Generally, the conidiogenesis is similar to that of other Blastosporae and Sympodulosporae. There is no specifically 'tretic' conidiogenesis: the presence of the pore surrounded by a melanized ring characterizes the subdivision Porosporae. However, there are intermediate forms between this group and the earlier ones:

— It is sometimes difficult to determine whether there is a pore or not.

— There is sometimes an annular thickening only on the conidiophore.

In spite of these difficulties, this distinction remains of great practical diagnostic value, and the Porosporae, all melanized, present common characters and seem like a family, and are fairly easy to identify.

The porospores are produced on conidial apparatus of varying development: percurrent or subapical proliferation, with fertile heads or nodes that may or may not be bulging, with conidia that are solitary or in chains, etc.

These conidia are unicellular, bicellular, phragmiate, or dictyate. In the phragmoconidia, the maturation may be acrospore or holospore (see Fig. I.7). The conidia in chains have a scar at the base and one or several at the tip. Finally, the characters of euseptation and distoseptation (Fig. I.8) have been found primarily in this group (Luttrell, 1963, 1964).

Some confusion has long existed about the definition of certain genera. In particular the phragmiate species and even dictyate species have generally been mixed in a large, rather miscellaneous genus, *Helminthosporium*. An important clarification was achieved with the segregation of the genera *Curvularia*, *Bipolaris*, *Drechslera*, and *Exserohilum* on the one hand (Alcorn, 1983; Ellis, 1966; Luttrell, 1969; Shoemaker, 1959; Sivanesan, 1985), and *Alternaria*, *Embellisia*, *Stemphylium*, and *Ulocladium* on the other hand (Simmons, 1967, 1969, 1971, 1983). *Helminthosporium*, *Exosporium*, *Spiropes*, *Corynespora*, and other genera seem to us to be still poorly defined, being heterogenous as regards several important characters, such as:

— septation (conidia euseptate or distoseptate);

— form of conidiophore and the conidiogenous cell in relation to the mode of proliferation;

— presence or absence of conidial chains.

These make identification difficult. The knowledge of teleomorphs may help in better classifying the anamorphs (Alcorn, 1983).

Key to identification of Porosporae

1	Conidia simple or phragmiate		2
	Conidia dictyate		25
2(1)	Amerospores or didymospores[1] (rarely 2-3 septa)		3
	Phragmospores		7
3(2)	Conidiophores micronematous (conidiogenous hyphopodites)	*Pirozynskia*	IX 26
	Conidiophores macronematous		4
4(3)	Amerospores (rarely 1-3 septa)	*Fusicladium*	IX 8
	Didymospores		5

[1]Certain *Spadicoides* (see 16, Fig. IX.7) have amerospores or didymospores.

5(4)	Conidia fusiform	*Fusicladiella*[2]	**IX 27**
	Conidia of different forms		6
6(5)	Conidia cylindrical, in chains	*Diplococcium*	**IX 6**
	Conidia pyriform isolated on conidiophores with twisted appearance	*Polythrincium*	**IX 28**
7(2)	Conidiophores spherical, micronematous	*Polyschema*	**IX 14**
	Conidiophores macronematous		8
8(7)	Conidia regularly in chains		9
	Conidia solitary (occasionally in chains)		10
9(8)	Conidiophores simple	*Corynesporopsis*	**IX 1**
	Conidiophores ramified	*Dendryphiella*	**IX 19**
10(8)	Conidia only apical (monotretic parent cell)		11
	Conidia apical and lateral (polytretic parent cell)		15
11(10)	Conidia variable, cylindrical, conical, club-shaped, on non-ramified conidiophores	*Corynespora*	**IX 2**
	Conidiophores ramified		12
12(11)	Regular, superposed verticils of compact conidiogenous cells	*Spondylocladiella*	**IX 24**
	Branches of different types		13
13(12)	Branches long		14
	Branches and conidiogenous cells compact	*Dendryphion*	**IX 18**
14(13)	Conidia uniformly coloured	*Dendryphiopsis*	**IX 12**
	Conidia discoloured	*Paradendryphiopsis*	**IX 13**
15(10)	Growths of conidiophore absent or recurrent		16
	Subterminal growths of conidiophore		17
16(15)	Conidia cylindrical with holospore maturation	*Spadicoides*	**IX 7**
	Conidia conical with acrospore maturation	*Helminthosporium*	**IX 3**
17(15)	Conidia euseptate, discoloured		18[3]
	Conidia distoseptate or euseptate, uniformly coloured		19
18(17)	Conidia biseptate with a dark, bulging median cell and small, light-coloured terminal cells	*Brachydesmiella*	**IX 25**
	Conidia with 3 or more septa, often curved and with more light-coloured cells at the extremities	*Curvularia*	**IX 17**
19(17)	Conidia rather symmetrical (fusoid, cylindrical)		20
	Conidia rather conical		24[4]

Continued on p. 158

---

[2]See also *Passalora* (Table VI.E 11).
[3]See also *Mycocentrospora* (Table VI.H 5).
[4]See also *Cercospora* (Table VI.E 9) and *Cercosporidium* (Table VI.E 10).

Table IX. Porosporae

Fig.	Genus	No. spp	Geographical distribution	Mode of life, substrates and impact	Teleomorphs	Additional references	Mol. biol.
1	Corynesporopsis	5	Eurasia, Cuba	Saprophyte, wood and litter			
2	Corynespora	35	Cosmopolitan	Saprophyte and parasite on plants	Corynesporasca		X
3	Helminthosporium	15	Cosmopolitan	Phytoparasite, lignicolous	Cochliobolus		X
4	Exosporium	17	Cosmopolitan	Saprophyte and parasite on plants			
5	Spiropes	28	Cosmopolitan	Mostly hyperparasite on Ascomycetes, and phytoparasite			
6	Diplococcium	10	Africa, N. Amer., Europe	Saprophyte on wood, bark, soil, fungicolous	Helminthosphaeria	Fol. G. Pr. 17: 295 (1982)	
7	Spadicoides	17	Cosmopolitan	Saprophyte on wood, bark, litter, fungicolous		Fol. G. Pr. 17: 295 (1982), Myc. Res. 95: 163 (1991)	
8	Fusicladium	40	Cosmopolitan	Generally a parasite on various organs of many plants	Acantharia, Apiosporina, Microcyclus, Venturia		
9	Pseudospiropes	26	Cosmopolitan	Mostly saprophyte on dead wood and plant debris	Melanomma, Strossmayeria	Mycot. 36: 383 (1990)	
10	Bipolaris	48	Cosmopolitan	Agent of brown spots on Poaceae, parasite mostly on plants, sometimes on humans and animals, hypovirulent	Cochliobolus	Myc. Pap. 158 (1987); CJB 76; 1558 (1998)	X
11	Exserohilum	20	Cosmopolitan, mostly tropical	Phytoparasite, bioherbicide	Setosphaeria	Myc. Pap. 158 (1987); CJB 76: 1558 (1998)	
12	Dendryphiopsis	4	Ubiquitous	Saprophyte, lignicolous	Kirschsteiniothelia		
13	Paradendryphiopsis	3	S. Africa, Europe	Saprophyte, lignicolous			
14	Polyschema	8	Africa, N. and S. Amer., India	Saprophyte, soil and plant debris			
15	Dictyopolyschema	1	Gr. Britain	Saprophyte, lignicolous			
16	Drechslera	24	Cosmopolitan	Graminicolous phytoparasite	Pyrenophora	Myc. Pap. 158 (1987); CJB 76: 1558 (1998)	X
17	Curvularia	36	Cosmopolitan	Graminicolous phytoparasite	Cochliobolus	Myc. Pap. 158 (1987)	X
18	Dendryphion	4	Cosmopolitan	Saprophyte and parasite on plants	Pleospora		
19	Dendryphiella	5	Cosmopolitan	Saprophyte in marine or terrestrial environment, 1 sp. parasite on leaves of Lantana		Myc. Res. 99: 771 (1995)	X
20	Stemphylium	20	Cosmopolitan	Saprophyte and phytoparasite	Pleospora		X
21	Alternaria	50	Cosmopolitan	Telluric, saprophyte or parasite of many plants, toxinogen	Comoclathris, Lewia	Rev. Myc. 29: 348 (1965); Mycot. 55: 55 (1995)	X
22	Embellisia	18	Cosmopolitan	Saprophyte, soil, etc., phytoparasite	Allewia	Mycot. 38: 251 (1990)	
23	Ulocladium	9	Cosmopolitan	Saprophyte, soil, etc., phytoparasite			
24	Spondylocladiella	1	N. Amer., Europe	Fungicolous			
25	Brachydesmiella	3	N. Amer., Europe, India	Saprophyte, lignicolous			
26	Pirozynskia	2	N. Amer.	On needles of Abies and Thuja, and leaves of Andromeda	Eupelte		
27	Fusicladiella	4	Eurasia	Phytoparasite			
28	Polythrincium	1	Cosmopolitan	Leaf pathogen on Trifolium	Cymadothea		

Porosporae

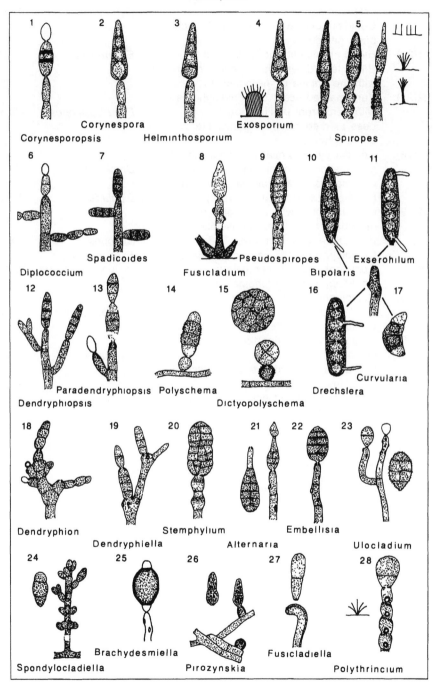

1 2 3 4 5

Corynespora
Corynesporopsis Helminthosporium Exosporium
Spiropes

6 7 8 9 10 11

Spadicoides Pseudospiropes Exserohilum
Diplococcium Fusicladium Bipolaris

12 13 14 15 16 17

Paradendryphiopsis Polyschema Curvularia
Dendryphiopsis Dictyopolyschema Drechslera

18 19 20 21 22 23

Dendryphion Stemphylium Embellisia
Dendryphiella Alternaria Ulocladium

24 25 26 27 28

Brachydesmiella Fusicladiella
Spondylocladiella Pirozynskia Polythrincium

20(19)	Conidiophores branched, conidiogenous cells compact	*Dendryphion*	**IX 18**
	Conidiophores with few or no branches, often geniculate		21
21(20)	Hilum prominent and melanized	*Exserohilum*	**IX 11**
	Hilum less differentiated, often integrated with the outline of the conidium		22
22(21)	Very marked scars on conidiophores with percurrent proliferation	*Pseudospiropes*	**IX 9**
	Less marked scars on geniculate conidiophores		23
23(22)	Conidia cylindro-oval with amphigenous germination (from any cell)	*Drechslera*	**IX 16**
	Conidia fusoid with bipolar germination (by the cells at the extremities)	*Bipolaris*	**IX 10**
24(19)	Conidiophores with sympodial, percurrent or subapical proliferation, isolated, fasciculate, or in coremia	*Spiropes*	**IX 5**
	Conidiophores grouped on stromata	*Exosporium*	**IX 4**
25(1)	Conidiophores micronematous	*Dictyopolyschema*	**IX 15**
	Conidiophores macronematous		26
26(25)	Conidiophores with percurrent proliferation	*Stemphylium*	**IX 20**
	Conidiophores with subterminal growth		27
27(26)	Conidia with tapering tips, often in chains	*Alternaria*	**IX 21**
	Conidia with rounded tips, sometimes in chains		28
28(27)	Septa and walls of equal thickness	*Ulocladium*	**IX 23**
	Thick transverse septa	*Embellisia*	**IX 22**

PHIALOSPORAE

In an attempt to set down the bases of a rational classification of Fungi Imperfecti, Vuillemin (1910a) defined a group in which 'the branch that serves as immediate support for the conidia often has the form of a flask with a venter and a neck ... which ... is called phialide (*phiala*, flask or bottle). The typical phialide has conidia formed exclusively at the end of the neck.'

In another publication, Vuillemin (1910b) specified that several conidia are formed at the top of the phialide. In an *Acremonium*, 'the conidium ... falls ... as soon as it is mature; but that does not exhaust its support, which soon produces a second conidium, and then still others.' In a *Spicaria* (presently *Paecilomyces*), 'the conidia are formed in a basipetal progression and in indefinite number from the tip of the phialide.' Later, the author specifies that these conidia are arranged in a chain and that 'the youngest conidium [is] at the tip of the phialide' (at the base of the chain). It is noted that this character of basipetal succession of conidia is insufficient to determine a phialide (there is also basipetal succession in the Meristem Arthrosporae, the Annellophorae, and the Annellidae). But it enables us to understand that a phialide can never produce ramified chains, unlike the basifugal chains of blastoconidia (e.g., *Cladosporium*). Therefore, if the phialoconidium has scars at the points at which it was attached to the preceding and following conidia, there can never be more than two scars, one basal, one apical (Fig. X.1).

At the Kananaskis symposium (Kendrick, 1971a), many specialists on Fungi Imperfecti proposed precise definitions for the different con-idiogenous organs. Their conception of the phialide is as follows: 'A coni-diogenous cell which produces, from a fixed conidiogenous locus, a basipetal succession of *enteroblastic* conidia, whose walls arise *de novo*.' It seems, therefore, that this modern definition of phialide, in comparison to that of Vuillemin, is only more precise in stating the enteroblastic nature of the conidiogenesis in the phialide. We will see what is involved in the production of the wall *de novo*.

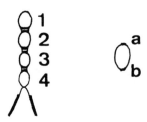

Fig. X.1. Basipetal succession of phialoconidia (the youngest is at the base). The isolated conidium has an apical scar (**a**) and a basal scar (**b**).

In sum, except for the possible case of phialides with solitary conidia (Gams, 1973), the phialide is a conidiogenous cell that buds from the same vegetative point into many conidia in a basipetal order. This is particularly visible in cases in which the conidia form chains (Fig. X.2).

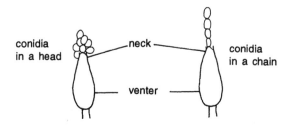

Fig. X.2. Arrangement of conidia on the phialide.

Various types of phialides are known (Fig. X.3).

We have already said in the introduction and in the discussion above that the Phialosporae form their conidia by a process of budding, which is **blastic**. On the other hand, the analysis that we present later of certain ultrastructural studies shows that the conidia (except the first) are of internal origin because they are limited by the single inner layer of the phialidic wall: the Phialosporae are **enteroblastic**, in contrast to Blastosporae, which are **holoblastic** (chapters III-IX).

Study of some examples of ultrastructure

Plain phialides with dry chains

The electron microscope studies are quite numerous, particularly for the genera *Aspergillus* and *Penicillium*. We will borrow from Roquebert (1981) the example of *Aspergillus tamarii* (Fig. X.4).

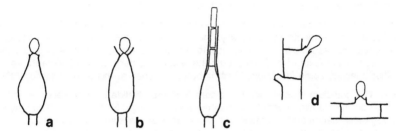

Fig. X.3. Types of phialides: **a:** plain phialide; the phialospores are produced from the top of the neck; **b:** phialide with collarette; a part of the wall of the phialide, flared out or more or less cylindrical, extends the neck; **c:** endophialide; a long, nearly cylindrical neck extends from the venter of the phialide. The meristematic point is generally located towards the top of the venter and several conidia may be present in the neck; **d:** all phialides do not have the typical form of a flask or a terminal position; there are examples of phialides with lateral neck.

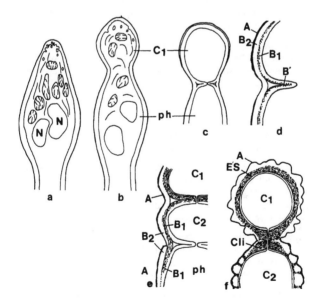

Fig. X.4. Stages in the conidiogenesis and maturation in the case of plain phialides with dry chains (*Aspergillus tamarii*): **a:** young phialide; **b:** primordium of conidial bud; **c-d:** individualization of the conidium by a septum; **e:** formation of a second conidium and septation; **f:** conidial maturation.

N: nucleus; C1: first conidium; C2: second conidium; ph: phialide; A: external film; B1, B2: inner and outer layers of the fundamental wall; B': invagination of the layer B1 at the level of the septum; ES: endospore; Cli: lamellar thickening.

a: The very young phialide has an apex similar to that of a growing hypha, with stratification and orientation of constituents; some vesicles are collected at the top (with probably some effect on parietal synthesis); closer to the base, the mitochondria and the endoplasmic reticulum are longitudinal; finally, there is the nucleus.

b: The apex bulges to form the primordium of the conidial bud; the vesicles and mitochondria distribute more equally; the endoplasmic reticulum becomes oriented in a concentric fashion; this corresponds to a halt in apical growth of a hyphal type. A nucleus migrates into the conidial bud; the wall of the neck of the phialide thickens. In other species, some longitudinal microtubules are seen at this stage (Hanlin, 1976).

c-d: The conidium thus formed is isolated from the phialide by a septation exactly the same as that known in vegetative hypha (chapter I, Fig. I.3); the fundamental layer B1 invaginates centripetally, forming two layers B' that are symmetrical in relation to the zone of cleavage (zc), leaving a pore at the centre. However, there are no Woronin bodies (even though the *Aspergillus* are anamorphs of Ascomycetes, as all the Phialosporae seem to be). The layers B' are continuous with B1 of the phialide and the spore.

The first conidium therefore has a wall in which all the layers, B1, B2, and A, are continuous with those of the phialide: **it is of holoblastic formation**, as in the (Holo-) Blastosporae.

e: The first conidium, C1, being formed and separated from the phialide by a septum, a second conidium, C2, is produced and begins to separate itself from the phialide. It is noted that its wall is formed only from the internal layer B1 of the phialide: it differentiates itself into internal layer B1 and external layers B2 and A, these last two ensuring the continuity and cohesion of the chain until maturity.

f: During the process of maturation, the conidium is surrounded by a thickened endospore. Between the two conidia, a lamellar thickness is formed, ensuring a fragile link between them. The external film A, separating from the endospore, constitutes the ornamentation of the mature conidium.

Plain phialides with mucous heads

Hammill (1974b) studied conidiogenesis in *Trichoderma saturnisporum*, in which the conidia form mucous heads (Fig. X.5).

There too, the first conidium is formed holoblastically, since its entire wall is primarily continuous with that of the phialide (Fig. X.5a). However, the outer layer B2, loosely fibrous, breaks up before maturity around the conidial primordium. The inner layer B1 of this is continuous with the fundamental inner layer of the phialide; it differentiates an electron-dense, mucilaginous layer A towards the exterior.

It must be noted that, from this stage, there is a thickening at the neck, made up of electron-dense fibres and located between the inner layer B1 and the outer layer B2. This differentiation of layers seems to be produced only towards the apex of the phialide. Under the thickening of the neck only a layer B can be distinguished.

The septation that isolates the conidia is produced at the narrowest part of the neck (Fig. X.5c). It is total, without pore, and the two halves of the septum that is cleaved will constitute the base of the conidium formed and the tip of the following conidium (Fig. X.5d). This will be surrounded only by the fundamental inner layer arising from the phialide, which is differentiated very quickly into a transparent inner layer and an electron-dense outer layer (Fig. X.5b-c).

In this species, the outer layer of the conidial wall separates in places from the inner layer, forming a sort of membranous appendage (cf. the ornamentations of the wall in *Aspergillus tamarii*). At the base it forms a fringe around the hilum, which it does not

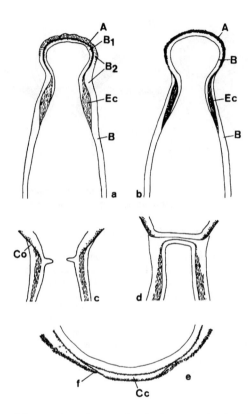

Fig. X.5. Stages of conidiogenesis and maturation in the case of plain phialides with mucous heads (*Trichoderma saturnisporum*): **a**: primordium of conidial bud of the first conidium; **b**: subsequent formation of the conidium; **c**: beginning of the septation between phialide and conidium; **d**: septation completed; **e**: base of the conidium. **A, B, B1, B2**: layers of the wall; **Ec**: thickening of the neck; **Co**: collarette; **f**: fringe around the hilum; **Cc**: scar layer of the hilum.

cover, but, at this place, the internal wall is thickened and has produced a supplementary layer (Cc) that thus scars the hilar region (Fig. X.5e).

The difference between dry conidia and mucous ones is not evident; it lies essentially, we believe, in the fact that the superficial film A is less condensed, more fibrous in the latter, which is characteristic of mucilages. On the other hand, the absence of chains seems to be linked to the absence of connections between conidia: there is no scar at the tip of the conidium, and at the base is a reduced hilum, scarred and invisible under light microscope.

Similarly, the small collarette (Co) that slightly extends the neck of the phialide beyond its narrowest zone, being close against the conidium that is forming (Fig. X.5c-d), cannot be seen under light microscope and the phialides of *Trichoderma* appear plain.

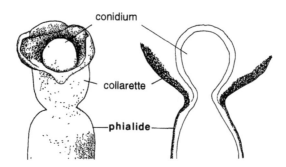

Fig. X.6. Phialides with collarette in *Phialophora* spp. (Reisinger, 1972; Cole and Samson, 1979) under SEM (at left), and in longitudinal section under TEM (right).

Phialides with collarette (Fig. X.6)

Although there can be plain phialides in *Phialophora richardsiae*, which are not different from those we have seen till now, this genus generally has phialides with a collarette. These result from the breaking up of the tip of the phialide wall after production of the first conidium. As in the preceding case, the conidium is formed holoblastically, since it is surrounded at the outset with all the layers of the parent cell, but finally it is enteroblastic, because it is surrounded at maturity by its own single wall issuing from the inner layer of the phialide. The separating septation appears at the narrow neck of the phialide, at the base of the collarette. According to Reisinger (1972), a rapid lysis of the external mucilaginous layers of the conidial wall generally prevents the observation of successive deposits of the collarette, but the latter is charged with granules of melanin.

Endophialides

Chalara elegans (Hawes and Beckett, 1977c): Fig. X.7a-d.

The phialides are long and have a slightly bulging basal venter and cylindrical neck. The conidia become individualized by septation towards the upper quarter or half of the phialide, in the neck.

Under transmission electron microscope (Fig. X.7d), the following observations can be made of the wall:
    — The venter seems to have a wall formed of a single layer B covered with a film A.
    — Above the venter a second inner layer begins to differentiate and thickens progressively.
    — Further up in the neck, this inner layer (Pi) becomes completely separated from the outer layer (Pe), and a mucilaginous material (Mu) appears between the two. An inner film forms in the neck of the phialide, which grows constantly towards the top, with its wall formed from constantly synthesized parietal material.
    — Still higher up, this filament fragments into more or less regular segments by total septation and then median cleavage. These segments are the conidia.

The zone demarcated between the point of differentiation of the inner layer, above the venter, and the point of separation of the inner and outer walls constitutes the meristem (M)

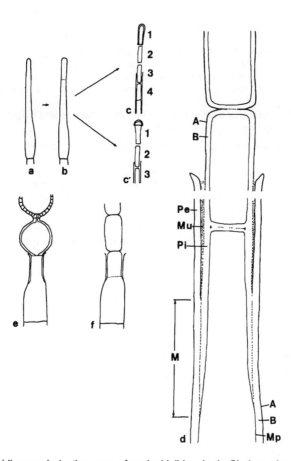

Fig. X.7. Conidiogenesis in the case of endophialides (a-d: *Chalara elegans*; e,f: *Chalara* anamorph of *Ceratocystis adiposa*). **a**: young phialide; **b**: delimitation of the first conidium; **c, c'**: various modes of rupture of the tip of the phialide, under the growth of the chain of conidia; **d**: details of an endophialide under TEM; **e**: spherical, melanized conidia produced in subapical position; **f**: subcylindrical, hyaline conidia produced deeper within the neck. **A, B**: layers of the wall; **M**: meristematic zone; **Mp**: plasmic membrane; **Mu**: mucilage; **Pe**: outer wall; **Pi**: inner wall.

in which the conidial wall is newly formed (as shown in the UV observation of phialides treated with Calcofluor).

The mucilaginous material between the inner and outer walls seems to serve the purpose of lubrication and then is transformed into layer A around the conidia at the time of their release from the neck of the phialide. These then form fragile chains.

The first conidium is slightly bulging at the tip and is either surrounded above and on the sides by the wall of the phialide (Hawes and Beckett, 1977c) (Fig. X.7c) or only surmounted with a small 'cap' (Ingold, 1981) (Fig. X.7c').

Chalara* anamorph of *Ceratocystis adiposa (Hawes and Beckett, 1977a, b; Hawes, 1979).

In this species, in which the phialide and especially the neck are much shorter than in *Chalara elegans*, the conidia are variable:

— When the zone of division is relatively deep inside the neck, fragile chains of subcylindrical conidia with thin walls, more or less hyaline, are produced (Fig. X.7f).

— When the conidia are delimited higher up, in a subapical position, they are spherical, a thick, melanized wall composed chiefly of a layer A continues between the conidia, a septal pore being closed by means of Woronin bodies, and the conidial chains are persistent (Fig. X.7e).

— Some intermediates exist in the form of conidia more or less bulging or 'pyriform', with a rounded tip and a cylindrical base.

These variations are linked essentially to the height of the septation zone in the neck of the phialide:

— If the conidial primordium is released from the phialide before septation, its wall is softer and it becomes rounded; also, it thickens and becomes melanized.

— If the conidial primordium is septate when it is still in the neck, the wall becomes rigid but remains thin and poorly melanized.

The more or less precocious septation between conidium and conidiogenous filament determines, therefore, the characteristics of conidia by determining those of the wall.

Conclusions

In all cases, there is a continuity between the inner layer of the phialidic wall and the wall of the conidial primordium. This inner layer is individualized more or less high up in the neck. The conidial primordia never appear like a protoplast (mass of protoplasm lacking a wall), as light microscope observations may lead one to think.

The conidial separation is schizolytic in all the Phialosporae studied. The lower half of the separating septum of a conidium will become part of the wall of the conidium that is later formed. The conidial wall is not, therefore, formed *de novo*.

The thickening often present at the neck of the phialide is not necessarily linked with the accumulation of residues of successive walls, since it is present even in young or monoconidial phialides. Such an accumulation may, however, be produced (phialoconidia in false chains or in heads). This thickening is absent in endogenous phialides, with the exception of *Termitaria*.

The first conidium is always formed in a holoblastic manner, that is, it is surrounded by parietal layers of the conidiogenous cell. When it is released, it may or may not carry the layers B2 and A of this conidiogenous cell: entirely in the case of plain phialides, partly in the endophialides, and not at all in phialides with a collarette.

The subsequent conidia are enteroblastic.

The analysis that we have made shows that the site of conidiogenesis is **fixed** in the Phialosporae.

We will see in chapter XI that in the Annellidae this site is displaced towards the top and that therefore the neck becomes slightly elongated with the production of each conidium. In *Cladobotryum* (Table X.B14), on the other hand, which we have put with the Phialosporae, this conidiogenous site recedes with the production of each conidium and the sporogenous cells progressively shorten, but they retain the appearance of a phialide (Minter et al., 1983a).

Key to identification of Phialosporae

| 1 | Hyphales | | 2 |
| 2 | Conidiomales | | 99 |

Hyphales _____

2(1)	Fungus hyaline		3
	Fungus more or less melanized		54
3(2)	Conidia unicellular		4
	Conidia variously septate		38

Hyaline ameroconidia _____

4(3)	Conidia in mucous heads		5
	Conidia in chains		26
5(4)	Phialides reduced and more or less mixed with fertile hyphae		6
	Phialides normally developed		11
6(5)	More or less levuriform, monilioid appearance		7
	Fungus mycelial		8
7(6)	Radulaspores, botryoblastospores	see *Aureobasidium*	(VIII 7)
	Phialides integrated or distinct	*Hormonema*	**C1**
8(6)	Phialides integrated with lateral collars on the hypha	*Cladorrhinum*	**C5**
	Phialides more or less developed		9
9(8)	Beaks more or less developed on the conidiophore	*Aphanocladium*	**A1**
	No such character		10
10(9)	Conidiophores in clumps with acropleurogenous phialospores	*Meria*	**A2**
	Phialides terminal, intercalary or lateral, more or less in brushes	*Sesquicillium*	**A3**
11(5)	Phialides isolated or irregularly grouped		12
	Phialides in bunches, verticils, heads, etc.		14
12(11)	Phialides elongated, conidia short		13
	Phialides bulbous, conidia highly elongated	*Harposporium*	**A4**

13(12)	Conidia globular to subcylindrical (on cords of hyphae more or less creeping = form sometimes wrongly called *Tilachlidium*, see E7)	*Acremonium*	**A5**
	Conidia ovoid with mucous rolls = hyphal form of the Stilbellaceae	*Hirsutella*	**E6**
14(11)	Phialides grouped on a terminal ampulla	*Gliocephalis*	**A6**
	Other mode of grouping		15
15(14)	Phialides in terminal brushes		16
	Phialides terminal and lateral		18
16(15)	Phialides arranged approximately at right angles on the conidiophores	*Trichoderma*	**A7**
	Phialides converging		17
17(16)	Acroblastosporous synanamorph	*Dimorphospora*	(VII A12)
	No such character, conidia aggregated in a common head with all the phialides in a brush	*Gliocladium*	**A8**
18(15)	Phialides in regular verticils		19
	Terminal and lateral phialides not in verticils		22
19(18)	Phialides and metulae very bulging	*Coryne*	**A9**
	No such character		20
20(19)	Dense heads of verticils at tip of conidiophore	*Gloiosphaera*	**A10**
	Verticils less dense		21
21(20)	Phialides fairly large, curved at tip	*Uncigera*	**A11**
	Phialides long and narrow	*Verticillium*	**A12**
22(18)	Conidiophores not differentiated, phialides conical or slightly bulging, conidia globular [See also the *Myrioconium* synanamorph of *Cristulariella* (Agonomycetes, XIII 17)]	hyphal stage of *Chaunopycnis*	**O1**
	No such characters		23
23(22)	Phialides generally implanted singly and curved	*Onychophora*	**A13**
	No such character		24
24(23)	Phialides with bulging base and narrow neck	*Tolypocladium*	**A14**
	Phialides in spindles		25
25(24)	Synanamorph with arched phragmoconidia	*Fusarium*	**B10**
	Only ameroconidia, phialides in dense apical bunches	*Harziella*	**A15**
26(4)	Phialides isolated		27
	Phialides grouped		29
27(26)	Conidiophores long	*Exochalara*	**A16**
	Phialides inserted laterally on the hyphae		28
28(27)	Phialides fusoid to cylindrical, chains of conidia with very pronounced connectives	*Sagenomella*	**A17**

	Phialides bulging very much between pedicel and tip	*Torulomyces*	**D3**
29(26)	Phialides grouped on a terminal bulge		30
	Phialides grouped in different ways		31
30(29)	Phialides produced simultaneously on a terminal ampulla of the conidiophore	*Aspergillus*	**A18**
	A phialide first produced at the tip, then others produced simultaneously	*Raperia*	**A19**
31(29)	Phialides in terminal brush		32
	Phialides terminal and lateral		36
32(31)	Masses of greenish and bluish conidia	*Penicillium*	**A20**
	Other colour		33
33(32)	The largest conidiophores with thick wall, spatulate or enlarged at the tip, metulae produced close together at the tip	*Merimbla*	**A21**
	Phialides exclusively in brushes		34
34(33)	All the elements of the conidiophore, including the phialides, with spiny wall; cylindrical phialides and conidia	*Geosmithia*	**A22**
	Non-spiny phialides		35
35(34)	Phialides and conidia non-cylindrical	*Penicillium*	**A20**
	Bulging phialides and cylindrical conidia	*Riclaretia*	**A23**
36(31)	Phialides often in terminal or lateral brushes sometimes isolated, conidia ovoid to fusoid	*Paecilomyces*	**A24***
	Less regular grouping		37
37(36)	Conidia pointed or rounded at the tip and truncated at the base	*Gabarnaudia*	**A25**
	Conidia truncated at both ends	*Septofusidium*	**B12**

Hyaline septate conidia

38(3)	Conidia isolated (aquatic fungi)		39
	Conidia in groups (terrestrial fungi)		40
39(38)	Phialides in penicilli, conidia with apical corona	*Heliscus*	**B1**
	Phialides apical and lateral, phragmoconidia arched	*Margaritispora*	**B2**
40(38)	Conidia in mucous heads		41
	Conidia in chains		52
41(40)	Phialides isolated or in small groups		42
	Phialides in groups		46
42(41)	Phialides lateral or reduced to a lateral collarette, didymo- or phragmospores	*Cyphellophora*	**B3**
	Phialides of normal size		43

---

*See also the phialidic synanamorph of *Diheterospora* (IV A 39).

43(42)	Macroconidia didymo- or phragmosporate, arched, microconidia amerosporate	*Fusarium*	**B10**
	A single type of conidium		44
44(43)	Long isolated phialides	*Cephalosporiopsis*	**B4**
	Phialides shorter and bulging, sometimes in small groups		45
45(44)	Conidia thin and curved at the ends	*Cylindrodendrum*	**B7**
	Conidia erect, thicker	*Exophiala*	**D13**
46(41)	Lateral brushes and sterile bristle terminating the conidiophore		47
	No such character		48
47(46)	Large didymo- or phragmoconidia in parallel bundles, septate bristle	*Cylindrocladium*	**B5**
	Small didymoconidia in spherical drops, aseptate bristle	*Cylindrocladiella*	**B6**
48(46)	Phialides curved	*Uncigera*	**A11**
	No such character		49
49(48)	Phialides short and bulbous, more or less isolated on the hypha or grouped in sporodochia, didymospores elongated	*Pseudomicrodochium*	**B8**
	Phialides of different forms and grouping		50
50(49)	Phialides verticillate	*Sibirina*	**B9**
	Phialides in more or less regular brushes		51
51(50)	Macroconidia didymo- or phragmosporate, erect or curved, with asymmetrical foot-cell, microconidia present	*Fusarium*	**B10**
	Like *Fusarium* but without asymmetrical foot-cell	*Cylindrocarpon*	**B11**
52(40)	Phialides long and narrow, isolated, conidia truncated at the ends	*Septofusidium*	**B12**
	Brushes of phialides		53
53(52)	Conidia subcylindrical	*Penicillifer*	**B13**
	Bulging conidia; 'retrophialides' diminishing in size with the production of conidia	*Cladobotryum*	**B14**

Melanized fungi

54(2)	Conidia unicellular	55
	Conidia variously septate	91

Ameroconidia

55(54)	Mucous heads	56
	Conidial chains	80
56(55)	Fungus levuriform, monilioid	57
	Fungus mycelial	58
57(56)	Radulaspores, botryoblastospores	see *Aureobasidium* (VIII 7)

	Phialides integrated or distinct	*Hormonema*	**C1**
58(56)	Phialides totally or partly integrated in the conidiophore, collars lateral		59
	Phialides developed normally		61
59(58)	Collarettes lateral on the parent-hypha, or basal, lateral or apical on the phialides	*Anthopsis*	**C2**
	Phialides acropleurogenous		60
60(59)	Conidiophores erect	*Zakatoshia*	**C4**
	Conidiophores circinate	*Cladorrhinum*	**C5**
61(58)	Phialides isolated or in small groups		62
	Phialides grouped		68
62(61)	Phialides long and narrow		63
	Phialides more or less bulging		64
63(62)	Hyphae and phialides melanized	*Pseudogliomastix*	**C6**
	Phialides melanized at the tip	*Gliomastix*	**C7**
64(62)	Phialides bulging between pedicel and apex	*Torulomyces*	**D3**
	No such character		65
65(64)	Phialides without collarettes	*Exophiala*	**D13**
	Collarette present		66
66(65)	Phialides long, often with regrowth	*Chloridium*	**C8**
	No such character		67
67(66)	Phialides normally apical and lateral at tip of conidiophore, sometimes isolated	*Phaeostalagmus*	**C10**
	Phialides isolated or more or less grouped on the parent hyphae	*Phialophora**	**C9**
68(61)	Phialides apical and/or lateral		69
	Phialides in regular apical bunches		76
69(68)	Phialides acropleurogenous on moniliform hyphae (fungi belonging to fumagines)	*Capnophialophora*	**C3**
	No such character		70
70(69)	Phialides apical and lateral at tip of conidiophores		71
	Presence of bristles or setiform conidiophores		72
71(70)	Phialides flask-shaped, with collarette	*Phaeostalagmus*	**C10**
	Phialides and conidia subcylindrical	*Stachylidium*	**D10**
72(70)	Phialides with curved tip	*Menispora*	**C11**
	Phialides more or less erect		73
73(72)	Phialophores terminated by an integrated phialide or polyphialide	*Dictyochaeta*	**C12**
	Phialides generally lateral on large setiform conidiophores that may or may not be ramified		74

---

*Dense groups, sometimes sporodochia. See *Cystodendron* (X F 16).

74(73)	Significant group of phialides forming a muff around the conidiophore	*Chaetopsina*	**C13**
	Groups of some phialides with different heights on the conidiophore		75
75(74)	Conidia spherical or oval	*Gonytrichum*	**C14**
	Conidia elongated, ciliate or not	*Codinaeopsis*	**C15**
76(68)	Phialides on apical ampulla		77
	Phialides in apical penicilli		78
77(76)	Phialides brown, without metula	*Custinghophora*	**C16**
	Bi-seriate heads of metulae and phialides	*Goidanichiella*	**C17**
78(76)	Penicilli uni-seriate (no metula)	*Stachybotrys*	**C18**
	Penicilli multi-seriate		79
79(78)	Penicilli with sterile bristles	*Gliocephalotrichum*	**C19**
	No bristles, long collarette	*Phialocephala*	**C20**
80(55)	Phialides isolated		81
	Phialides in groups		87
81(80)	Endophialides		82
	Phialides plain or with collarettes		83
82(81)	Phialides single, with or without synanamorph	*Chalara*	**D1**
	Phialides with sterile bristles	*Chaetochalara*	**D5**
83(81)	Conidia cuneiform, capitate hyphae	*Catenularia*	**D2**
	No such characters		84
84(83)	Phialides bulging between foot and apex	*Torulomyces*	**D3**
	Phialides long and narrow		85
85(84)	Chains persistent with connectives	*Sagenomella*	**A17**
	Chains fragile		86
86(85)	Melanization at the level of phialide neck	*Gliomastix*	**C7**
	No such character = see isolated forms of *Paecilomyces*		**A24**
87(80)	Apical ampullae with or without metulae	*Aspergillus*	**A18**
	Different grouping		88
88(87)	Phialides apical and lateral	*Acrophialophora**	**D4**
	Penicilli terminal		89
89(88)	With metulae	*Thysanophora*	**D7**
	Without metulae		90
90(89)	Penicilli diverging	*Eladia*	**D8**
	Penicilli converging (previously *Memnoniella*)	*Stachybotrys*	**C18**

Septate conidia

91(54)	Endophialides		92
	Exophialides plain or with collarettes		95

*See also *Paecilomyces*, from which *Acrophialophora* is distinguished by its warty conidiophores.

92(91)	Conidia hyaline or subhyaline		93
	Conidia melanized		94
93(92)	Phialides with a venter and a neck	*Chalara*	**D1**
	Phialides short, cylindrical, producing phragmo- or dictyospores	*Ascoconidium*	**F23**
94(92)	Phialides more or less flared, producing cuneiform conidia at the tip	*Sporoschismopsis*	**D11**
	Endophialides, cylindrical conidia, capitate hyphae	*Sporoschisma*	**D12**
95(91)	Fungus levuriform	*Hormonema*	**C1**
	Fungus filamentous		96
96(95)	Conidia phragmiate, curved, arranged in staggered rows in short chains	*Fusariella*	**D9**
	Conidia in mucous heads		97
97(96)	Small phialides, more or less isolated	*Exophiala*	**D13**
	Polyphialides on or with large bristles		98
98(97)	Sterile bristles and separated polyphialides	*Cylindrotrichum*	**D6**
	Conidiophores setiform with verticils of polyphialides	*Chaetopsis*	**D14**

Conidiomales

99(1)	Conidioma of 'coremium' type		100
	No such character		136

With coremia

100(99)	Coremia crude (cords of hyphae)	see *Acremonium* (A5) *Gliomastix* (C7)	
	Coremia characteristic		101
101(100)	Conidia endogenous, unicellular, hyaline		102
	Conidia exogenous, uni- to multicellular, hyaline or melanized		103
102(101)	Coremia determinate	*Endosporostilbe*	**E19**
	Coremia indeterminate	*Chalarodendron*	**E20**
103(101)	Conidia unicellular		104
	Conidia multicellular		128
104(103)	Fungus entomophagous		105
	Fungus non-entomophagous		110
105(104)	Phialides produced on metulae		106
	No such character		107
106(105)	Phialides thickened at the apex	*Gibellula*	**E12**
	Phialides not thickened at the apex	*Tilachlidiopsis**	**E3**
107(105)	Coremia indeterminate	*Tilachlidium*	**E7**
	Coremia determinate, with highly mucoid capitulum		108

---

*Tilachlidiopsis, previously considered a Phialidae (Morris, 1963), has been demonstrated to be an Arthrosporae (Stalpers et al., 1991), cf. chapter II.

108(107) Conidium 3 µm, yellow to orange,
in mass   *Polycephalomyces*   **E5**
Conidium > 5 µm, red to black, in mass   109
109(108) Conidium fusiform, hyaline to light brown   *Synnematium*   **E11**
Conidium oblong, subcylindrical, hyaline   *Hirsutella*   **E6**
110(104) Coremia intensely pigmented   111
Coremia pale   117
111(110) Conidium fusiform, ratio of l/w > 2   112
Conidium globular or ellipsoid,
ratio l/w < 2   113
112(111) Capitulum with lateral bristles   *Phaeostilbella*   **E14**
Capitulum with radiating bristles   *Saccardaea*   **E16**
113(111) Stipe with ornamental cells, globular to
ellipsoid   *Stilbella*   **E2**
Stipe without ornamental cells   114
114(113) Stipe on a developed basal stroma   *Stromatographium*   **E17**
Basal stroma absent   115
115(114) Length of conidia > 10 µm   *Didymostilbe*   **E8**
Length of conidia < 10 µm   116
116(115) Coremia spindle-shaped, with
monomitic stipe   *Phialographium*   **E15**
Coremia thick with dimitic stipe   *Crinula*   **E13**
117(110) Conidia with bristles   *Thozetella*   **F17**
Conidia muticate   118
118(117) Length of conidia > 10 µm   *Didymostilbe*   **E8**
Length of conidia < 10 µm   119
119(118) On resin of conifers   *Eustilbum*   **E4**
On other substrates   120
120(119) Coremia gelatinous   121
Coremia fleshy or fibrous, non-gelatinous   122
121(120) Conidia allantoid, coremia robust, highly
gelatinous   *Coryne*   **A9**
Conidia oblong, cylindrical, coremia
spindly and not very gelatinous   *Dendrostilbella*   **E1**
122(120) Coremia indeterminate   *Tilachlidium*   **E7**
Coremia determinate   123
123(122) Conidiophore highly ramified, frequently
on 3-4 verticils   *Gliocladium*   **A8**
No such character   124
124(123) Coremium brown-red, turning blood red
in KOH   *Tubercularia*   **F3**
No such character   125
125(124) Coremium having long bristles with
thick wall   126
Coremium lacking bristles with thick wall   127
126(125) Conidia in a mass, slightly coloured   *Volutella*   **F4**
Conidia in a mass, dark green to black   *Myrothecium*   **F9**
127(125) Conidia melanized   *Myrothecium*   **F9**
Conidia more or less hyaline   *Stilbella*   **E2**

128(103)	Conidia uniseptate	129
	Conidia multiseptate	134
129(128)	Coremia deeply coloured	130
	Coremia pale	131
130(129)	On rhizomorphs of *Armillaria mellea*	*Pseudographiella* **E18**
	Matrix other than fungal	*Didymostilbe* **E8**
131(129)	Coremia with marginal, highly warty hypha	*Actinostilbe* **F12**
	Ornamentations of marginal hyphae, if present, are not marked and are limited to subapical region of stipe	132
132(131)	Polyphialides present	*Stachycoremium* **E9**
	Polyphialides absent	133
133(132)	Conidia clavate to fusiform with thin wall	*Stilbella* **E2**
	Conidia ellipsoid with thick wall	*Didymostilbe* **E8**
134(128)	Conidia uni- to triseptate, fusiform	*Pseudographiella* **E18**
	Conidia multiseptate, fusiform to falciform	135
135(134)	Conidia with asymmetrical foot-cell	*Fusarium* **B10**
	Conidia with non-pediform base	*Atractium* **E10**
136(99)	Conidioma of sporodochium type	137
	Other type of conidioma	161

With sporodochia

137(136)	Sporodochium made up of large, single-layered cells	*Mycoleptodiscus* **F22**
	No such character	138
138(137)	Sporodochium with hyaline basal stroma	*Pseudomicrodochium* **B8**
	No such character	139
139(138)	Conidia endogenous	140
	Conidia exogenous	144
140(139)	Phialides melanized	141
	Phialides hyaline	143
141(140)	Conidia multicellular	*Ascoconidium* **F23**
	Conidia unicellular	142
142(141)	Conidia spherical	*Cystodendron* **F16**
	Conidia square	*Bloxamia* **F20**
143(140)	Conidia ovoid, not truncated at extremities	*Endoconidium* **F5**
	Conidia truncated at extremities	*Endoconospora* **F6**
144(139)	Conidia more or less coloured	145
	Conidia hyaline	147
145(144)	Conidia catenate	*Metarhizium* **F10**
	Conidia non-catenate	146
146(145)	Conidiophores ramified, phialides hyaline	*Myrothecium* **F9**
	No conidiophore, phialides melanized	*Stromatocrea* **F21**
147(144)	Conidia unicellular	148

Conidia multicellular		157
148(147) Sporodochia with sterile bristles		149
Sporodochia lacking sterile bristles		151
149(148) Bristles non-melanized	*Volutella*	**F4**
Bristles melanized		150
150(149) Bristles smooth	*Vermiculariopsiella*	**F18**
Bristles asperulate	*Sarcopodium*	**F2**
151(148) Conidia mixed with sterile hooked elements, warty towards the tip	*Thozetella*	**F17**
No such character		152
152(151) Conidia acicular, filiform	*Kmetia*	**F1**
Conidia non-acicular		153
153(152) Sporodochia parasitizing sori of 'rust'	*Tuberculina*	**F8**
No such character		154
154(153) No conidiophores, conidia with thick wall	*Phialophoropsis*	**F7**
Presence of conidiophores, conidia with thin wall		155
155(154) Phialides terminal grouped in penicilli	*Dendrodochium*	**F11**
Phialides acro-pleurogenous		156
156(155) Phialides bulbous	*Agyriella*	**F19**
Phialides elongated, cylindrical	*Tubercularia*	**F3**
157(147) Conidia didymosporate	*Actinostilbe*	**F12**
Conidia phragmosporate or helicosporate		158
158(157) Conidia spiral	*Vanbeverwijkia*	**F15**
Conidia non-helicoid		159
159(158) Conidia with apical protuberances	*Heliscus*	**B1**
Conidia without protuberances at the tip (often mixed with unicellular microconidia)		160
160(159) Macroconidia obtuse at both ends	*Cylindrocarpon*	**F13**
Macroconidia acute at the tip and with pediform base	*Fusarium*	**F14**
161(136) Conidioma acervular (sometimes close to the sporodochium or discoid cupuloid)		162
Conidioma pseudopycnidial or pycnidial		200

With acervuli and cupules

162(161) Conidioma acervular		163
Conidioma discoid cupuloid		191
163(162) Conidioma aristate		164
Conidioma muticate		166
164(163) Conidia 0-septate with sometimes an apical appendage	*Colletotrichum*	**G1**
Conidia 0- to uniseptate, always with bristles		165
165(164) A bristle at each end, conidia hyaline	*Pseudolachnea*	**H13**
An apical bristle, several basal bristles, conidia hyaline to brown	*Polynema*	**G2**

166(163)	Conidia brown	167
	Conidia hyaline	168
167(166)	Conidia unicellular, light brown *Greeneria*	**G3**
	Conidia dictyate, dark brown *Kaleidosporium*	**G4**
168(166)	Conidia septate	169
	Conidia non-septate	174
169(168)	Conidia uniseptate	170
	Conidia 1-3-septate	172
170(169)	Conidia short *Marssonina*	(V D1)
	Conidia long	171
171(170)	Conidia without interconidial anastomoses, with pediform base *Pycnofusarium*	**G5**
	Interconidial anastomoses, conidia non-pediform *Titaeospora*	**G6**
172(169)	Conidia acrogenous	173
	Conidia acropleurogenous *Myriellina*	**H4**
173(172)	Acervulus on an huge intramatrical stroma *Cheilaria*	**H1**
	No such character *Septogloeum*	**G7**
174(168)	Conidia having an apical appendage	175
	Conidia lacking appendages	176
175(174)	Apical appendage cellular, vermiform *Rhabdogloeum*	**G8**
	Apical appendage mucous, infundibuliform *Rhodesiopsis*	**G9**
176(174)	Conidioma intermediate between acervulus and sporodochium	177
	Conidioma typically acervular	179
177(176)	Fertile surfaces on either side of a sterile central stroma *Discosporina*	**H2**
	No such character	178
178(177)	Conidia cylindrical, thin conidiogenous cells *Cytogloeum*	**G10**
	Conidia ellipsoid, compact conidiogenous cells *Kabatina*	**H3**
179(176)	Conidiogenous cells integrated	180
	Conidiogenous cells distinct, or a mixture of distinct and integrated	181
180(179)	Conidiogenous cells long and cylindrical *Phacostroma*	**H6**
	Conidiogenous cells short, bulbous *Gloeosporidiella*	**G11**
181(179)	Conidiogenous cells distinct	182
	Conidiogenous cells a mixture of distinct and integrated	186
182(181)	Conidia pyriform, obpyriform or turbinate	183
	Conidia not pyriform or turbinate	184
183(182)	Conidia pyriform to obpyriform *Monostichella*	**G12**
	Conidia turbinate *Cryptocline*	(XI B1)
184(182)	Conidia with clearly visible hilum *Cryptosporiopsis*	(XI B2)
	Conidia with hilum barely or not visible	185
185(184)	Acervulus subcuticular *Asteroma*	**G13**
	Acervulus intraepidermal *Cylindrogloeum*	**G14**

186(181) Conidia more or less guttulate   *Rhodesia*   **G15**
        Conidia not guttulate   187
187(186) Length of conidia < 10 μm   188
        Length of conidia > 10 μm or intermediate   189
188(187) Conidia pyriform   *Gloeosporidina*   **G16**
        Conidia ellipsoid oval   *Sphaceloma*   **G17**
189(187) Length of conidia > 10 μm   190
        Conidium regular cylindrical, length
        intermediate   *Cylindrosporium*   **G18**
190(189) Conidia irregular cylindrical   *Discogloeum*   **G19**
        Conidia ellipsoid to pyriform oval   *Discula*   **G20**
191(162) Conidioma muticate   192
        Conidioma aristate   195
192(191) Conidioma discoid, conidia longer
        than 10 μm   193
        Conidioma cupuloid, conidia shorter
        than 10 μm   194
193(192) Conidia unicellular seleniform   *Pseudostegia*   **H5**
        Conidia multicellular filiform   *Septopatella*   (VI G9)
194(192) Conidia spherical   *Acleistia*   **H7**
        Conidia ellipsoid oval   *Agyriellopsis*   **H8**
195(191) Bristles hyaline   *Hainesia*   **H11**
        Bristles brown   196
196(195) Conidia brown   *Hoehneliella*   **H12**
        Conidia hyaline   197
197(196) Conidia shorter than 10 μm   *Dinemasporium*   **H9**
        Conidia longer than 10 μm   198
198(197) Conidia unicellular   *Stauronema*   **H14**
        Conidia multicellular   199
199(198) Conidia cylindrical with 2-3 bristles
        at each end   *Dwayalomella*   **H10**
        Conidia curved, fusoid, with 1 bristle
        at each end   *Pseudolachnea*   **H13**
200(161) Conidioma pseudopycnidial   201
        Conidioma pycnidial   303

Pseudopycnidia

201(200) Conidioma ramified   *Conidioxyphium*   **I 1**
        Conidioma non-ramified   202
202(201) Pseudopycnidia simple, often monolocular   203
        Pseudopycnidia complex, multilocular   248
203(202) Pseudopycnidia flat or columnar   204
        Pseudopycnidia not flat or columnar   225
204(203) Pseudopycnidia flat   205
        Pseudopycnidia columnar   221
205(204) Conidioma is typically a thyriopycnidium   206
        No such character   209
206(205) Conidia unicellular   207
        Conidia multicellular   *Rhizothyrium*   **I 14**

207(206)	Conidia filiform, often curved	*Actinothyrium*	I 13
	Conidia non-filiform		208
208(207)	Conidia without appendages	*Tubakia*	I 16
	Conidia with appendage at each end	*Tracylla*	I 15
209(205)	Conidia hyaline, unicellular		210
	Conidia multicellular		220
210(209)	Presence of hyphae with hyphopodia	*Peltasterinostroma*	J10
	Absence of hyphopodia		211
211(210)	Conidia with apical appendage	*Diachorella*	I 3
	Conidia without appendages		212
212(211)	Paraphyses present	*Coleophoma*	I 2
	Paraphyses absent		213
213(212)	Conidia acropleurogenous		214
	Conidia acrogenous		215
214(213)	Conidiogenous cells without collarette	*Pilidium*	I 8
	Conidiogenous cells with collarette	*Siroplacodium*	I 10
215(213)	Conidioma joined with its ascoma	*Leptothyrium*	I 11
	No such character		216
216(215)	Conidia longer than 15 µm		217
	Conidia shorter than 15 µm		218
217(216)	Conidia arched, longer than 30 µm	*Cryptosporium*	(XI B11)
	Conidia variously curved, shorter than 30 µm	*Apomelasmia*	K9
218(216)	Conidia fusiform, straight, not guttulate	*Pseudothyrium*	I 9
	Conidia guttulate		219
219(218)	Conidia oval-elliptic	*Myxothyrium*	I 5
	Conidia cylindrical	*Xyloglyphis*	K12
220(209)	Conidia hyaline, arched, without appendages	*Periperidium*	I 6
	Conidia hyaline, straight, with an apical appendage	*Uniseta*	(V C11)
221(204)	Conidia unicellular		222
	Conidia multicellular, often arched		224
222(221)	Conidia oval elliptic to cylindrical, without appendage		223
	Conidia lenticular fusiform, with appendages	*Eleutheromyces*	K5
223(222)	Conidioma of *textura porrecta* without ostiolar bristles	*Cornucopiella*	K2
	Conidioma of *textura angularis* with ostiolar bristles	*Cylindroxyphium*	K3
224(221)	Conidia bicellular, acrogenous	*Corniculariella*	K1
	Conidia tetracellular, pleurogenous	*Sphaerographium*	K6
225(203)	Conidioma superficial		226
	Conidioma erumpent or intramatrical		229
226(225)	Conidioma with dark bristles	*Amerosporium*	J1
	Conidioma glabrous on the surface		227
227(226)	Conidioma of gelatinous consistency	*Gelatinopycnis*	K11
	Conidioma of carbonaceous consistency		228

228(227) Conidia short, unicellular   *Phacidiopycnis*   (XI B9)
          Conidia long, multicellular   *Pocillopycnis*   I 18
229(225) Conidioma erumpent   230
          Conidioma intramatrical   233
230(229) Conidiogenous cells percurrent, conidia
          0- to 1-septate, conidioma parasite on
          sori of Uredinales   *Sphaerellopsis*   (XI B12)
          No such characters   231
231(230) Conidioma highly stromatic, ostiolate   *Dothichiza*   **K14**
          Conidioma not highly stromatic,
          non-ostiolate   232
232(231) Conidia ellipsoid, erect, guttulate   *Paradiscula*   **J9**
          Conidia fusoid, curved, non-guttulate   *Phlyctema*   I 17
233(229) Conidioma ostiolate   234
          Conidioma non-ostiolate   239
234(233) Ostiole distinctly rostrate   *Cytonaema*   **K4**
          Ostiole not rostrate   235
235(234) Conidioma buried in a pseudostroma
          of *textura intricata*   *Neoplaconema*   **J8**
          No such character   236
236(235) Conidia hyaline   237
          Conidia brown   300
237(236) Conidioma hyperparasite of the genus
          Phyllachora   *Davisiella*   **J4**
          No such character   238
238(237) Conidia cylindrical, 0- to 3-septate,
          with appendages   *Dilophospora*   **J5**
          Conidia ellipsoid to fusiform, 0-septate,
          without appendages   *Phomopsis*   I 7
239(233) Conidioma linear   *Sphaeriothyrium*   **J14**
          Conidioma more or less circular, not
          linear   240
240(239) Conidia small, < 10 μm long   241
          Conidia large, > 10 μm long   243
241(240) Conidia turbinate, in acropetal chain   *Sirexcipula*   I 19
          No such characters   242
242(241) Upper wall of conidioma thin, uniformly
          coloured   *Siroplacodium*   I 10
          Upper wall of conidioma thick and darker   *Sporonema*   **J15**
243(240) Conidia bicellular   *Helhonia*   **J6**
          Conidia monocellular   244
244(243) Conidiogenous cell polyphialidic   *Sarcophoma*   **J13**
          Conidiogenous cell monophialidic   245
245(244) Conidiogenous cell percurrent   *Discosporium*   (XI B10)
          Conidiogenous cell non-percurrent   246
246(245) Conidia oval ellipsoid, sometimes lobed,
          often guttulate   *Phacidiopycnis*   (XI B9)
          Conidia fusiform, non-guttulate   247

247(246) Conidioma brownish-yellow, conidia
      curved   *Phlyctema*   I 17
           Conidioma dark brown, conidia erect or
      slightly curved   *Cryptomycella*   K10
248(202) Pseudopycnidium not pulvinate   249
           Pseudopycnidium pulvinate   278
249(248) Opening by an ostiole   250
           No ostiole, opening by tearing   265
250(249) Wall of conidioma of *textura intricata*   *Microdiscula*   L17
           Wall of conidioma predominantly of
      textura angularis   251
251(250) Paraphyses present   252
           Paraphyses absent   253
252(251) Conidia brown, monocellular   *Phaeocytostroma*   J11
           Conidia brown, multicellular   *Massariothea*   J7
253(251) Conidiogenous cells cylindrical   254
           Conidiogenous cells ampulliform, doliiform   261
254(253) Conidiogenous cells percurrent, conidia
           dark and multicellular, microconidia
      hyaline and unicellular   *Hendersoniopsis* (XI B13)
           No such characters   255
255(254) Conidia with appendages   256
           Conidia without appendages   258
256(255) Appendages mucilaginous, infundibuliform   *Ceuthospora*   L8
           Appendages cellular   257
257(256) Conidia tricellular, appendage apical   *Chaetoconis*   L9
           Conidia unicellular, appendage basal   *Strasseria*   M6
258(255) Conidia allantoid   *Cytospora*   L11
           Conidia ellipsoid to fusiform   259
259(258) Ostiole papillate and mucilaginous   *Dothichiza*   K14
           No such character   260
260(259) Conidia fusiform, < 15 μm long, often
           mixed with filiform conidia more or less
      curved   *Phomopsis*   I 7
           Conidia fusiform, > 15 μm long,
      sometimes mixed
      with bacilliform microconidia   *Fusicoccum*
                                                 (IV E10; V C13)
261(253) Conidia with appendages   *Dilophospora*   J5
           Conidia without appendages   262
262(261) Conidia hyaline   263
           Conidia brown   300
263(262) Conidia multicellular, conidioma linear   *Cytoplacosphaeria*   L10
           Conidia uni- to bicellular   264
264(263) Conidioma more or less irregular to linear,
      pseudostromatic   *Placonemina*   L19
           Conidioma globular, stromatic   *Ceuthodiplospora*   L7
265(249) Conidioma forming a black crust on
      leaves or under bark of maple   266

	No such character		267
266(265)	Tarry crust on leaves, conidia hyaline	*Melasmia*	I 4
	Powdery crust under bark, conidia light brown	*Cryptostroma*	**M8**
267(265)	Conidioma of *textura angularis*		270
	Conidioma of different texture		268
268(267)	Conidioma of *textura intricata* joined with its ascoma	*Amphicytostroma*	**M7**
	Conidioma of *textura globulosa*		269
269(268)	Conidiogenous cells cylindrical	*Pleurocytospora*	**L20**
	Conidiogenous cells ampulliform, doliiform	*Podoplaconema*	**M2**
270(267)	Conidia brown		271
	Conidia hyaline		273
271(270)	Paraphyses present	*Phaeocytostroma*	**J11**
	Paraphyses absent		272
272(271)	Conidiogenous cells cylindrical, often percurrent	*Cyclothyrium*	(XI B15)
	Conidiogenous cells ampulliform or short cylinders, not percurrent	*Cytoplea*	**M9**
273(270)	Conidia with mucous appendages	*Allantophomopsis*	**L1**
	Conidia without appendages		274
274(273)	Conidia large, > 10 µm long		275
	Conidia small, < 10 µm long		276
275(274)	Conidia 0- to 2-septate, not guttulate	*Diplodina*	**L13**
	Conidia 0-septate with one large central drop	*Cyclodomus*	**K13**
276(274)	Conidiophores ramified, septate	*Pleuroplaconema*	**M1**
	Conidiophores absent		277
277(276)	Conidia cylindrical, non-guttulate	*Asteromellopsis*	**L3**
	Conidia ellipsoid, attenuated at the base, some guttulate	*Sclerophoma*	**M3**
278(248)	Conidioma ostiolate		279
	Conidioma not ostiolate, dehiscence by rupture		284
279(278)	Conidioma red		280
	Conidioma dark brown to black		281
280(279)	Ostiole rostrate, conidia ellipsoid	*Endothiella*	**M10**
	Ostiole non-rostrate, conidia bacilliform	*Zythiostroma*	**J16**
281(279)	Wall of only *textura oblita* or mixed		282
	Wall of only *textura angularis* or mixed		283
282(281)	*Textura oblita*, conidia 0-septate	*Dothiorina*	**L14**
	Textura oblita outside, *textura intricata* inside, conidia 3-septate	*Topospora*	**K8**
283(281)	Wall of *textura angularis*, conidia without appendages	*Camaropycnis*	**L6**
	Two types of texture: *angularis* and *intricata*, conidia with a basal appendage	*Strasseriopsis*	**K16**
284(278)	Dehiscence by longitudinal fissure	*Paradiscula*	**J9**

	No such character		285
285(284)	Conidioma of homogeneous texture		286
	Conidioma of heterogeneous texture		297
286(285)	Conidioma of *textura intricata* or *oblita*		287
	Conidioma of *textura angularis*		288
287(286)	Conidioma of *textura intricata*	*Aschersonia*	**L2**
	Conidioma of *textura oblita*	*Gelatinosporium*	**K15**
288(286)	Conidia septate		289
	Conidia aseptate		291
289(288)	Conidia bicellular	*Sirococcus*	**M5**
	Conidia multicellular		290
290(289)	Conidia phragmiate, hyaline	*Brunchorstia*	**L5**
	Conidia dictyate, brown	*Stegonsporiopsis*	**K7**
291(288)	Conidia large, > 10 μm long		292
	Conidia small, < 10 μm long		293
292(291)	Conidia cylindrical	*Blennoria*	**L4**
	Conidia oval ellipsoid, sometimes of lobate contour	*Phacidiopycnis*	(XI B9)
293(291)	Conidiophores absent	*Sclerophoma*	**M3**
	Conidiophores present		294
294(293)	Conidiogenous cells polyphialidic	*Pragmopycnis*	**J12**
	Conidiogenous cells monophialidic		295
295(294)	Conidia cylindrical	*Sirodothis*	**M12**
	Conidia oblong elliptic to pyriform		296
296(295)	Conidia acropleurogenous	*Gyrostroma*	**L16**
	Conidia acrogenous	*Scleropycnis*	**M4**
297(285)	Conidia aseptate, small	*Cytosporella*	**L12**
	Conidia septate		298
298(297)	Conidia bicellular, sometimes tetracellular but not fusarioid		299
	Conidia tetracellular, fusarioid	*Botryocrea*	**J2**
299(298)	Conidia filiform, curved or sigmoid	*Foveostroma*	**I 12**
	Conidia ellipsoid	*Fuckelia*	**M11**
300(236	Conidia multicellular, non-dictyate		301
and 262)	Conidia multicellular, dictyate	*Camarographium*	**J3**
301(300)	Conidia verruculose, paraphyses absent	*Endocoryneum*	**L15**
	Conidia smooth		302
302(301)	Paraphyses present	*Massariothea*	**J7**
	Paraphyses absent, *Scytalidium* stage frequently present	*Nattrassia*	**L18**

Pycnidia

303(200)	Pycnidium atypical, formed of loose aerial hyphae, interlaced, delimiting a glomerular body more or less spherical	*Chaunopycnis*	**O1**
	Typical pycnidium without these characters		304
304(303)	Pycnidium with bristles (aristate)		305
	Pycnidium without bristles (muticate)		313

305(304)	Bristles hyaline	*Angiopomopsis*	(XI B7)
	Bristles brown		306
306(305)	Bristles generally aseptate	*Chaetodiplodia*	**N1**
	Bristles always septate		307
307(306)	Pycnidium of *textura prismatica*	*Chaetosphaeronema*	**N2**
	Pycnidium of *textura angularis*		308
308(307)	Pycnidium ostiolate, rostrate	*Wojnowicia*	**N3**
	Pycnidium ostiolate, not rostrate		309
309(308)	Ostiole linear	*Chaetomella*	**N4**
	Ostiole circular		310
310(309)	Conidiophores filiform, ramified, conidia acropleurogenous	*Pyrenochaeta*	**N5**
	No conidiophores, conidia acrogenous		311
311(310)	Conidia long, > 40 μm long, septate	*Chaetoseptoria*	**N6**
	Conidia short, < 40 μm long, septate or not septate		312
312(311)	Conidia cylindrical, sometimes septate	*Chaetosticta*	**N7**
	Conidia ellipsoid, non-septate	*Phoma*	**N8**
313(304)	Hyperparasite of Erysiphales	*Ampelomyces*	**N9**
	No such character		314
314(313)	Pycnidium rostrate		315
	Pycnidium not rostrate		319
315(314)	Conidia dark		316
	Conidia hyaline		317
316(315)	Conidia dictyate, with a mucoid appendage at each extremity	*Amarenographium*	**N10**
	Conidia phragmiate, without appendage	*Ceratopycnis*	**N11**
317(315)	Conidia without appendages	*Metazythiopsis*	**N12**
	Conidia with appendages		318
318(317)	Appendages cellular	*Eleutheromycella*	**N13**
	Appendages mucoid	*Choanatiara*	**N14**
319(314)	Pycnidium brilliantly coloured, of *textura intricata*		320
	Pycnidium golden brown to dark brown		321
320(319)	Conidiophore filiform, ramified	*Pycnidiella*	**N15**
	Conidiophore filiform, not ramified	*Rhodosticta*	**N16**
321(319)	Pycnidium of homogeneous texture		326
	Pycnidium of heterogeneous texture		322
322(321)	Texture *intricata-angularis*		323
	Texture *globulosa-angularis*		325
323(322)	Base of the pycnidium *intricata*, tip *angularis*	*Phacostromella*	**N17**
	Exterior of pycnidium *intricata*, interior *angularis*		324
324(323)	Conidia globular, non-septate, brown	*Epithyrium*	**N18**
	Conidia cylindrical, distoseptate, brown	*Massariothea*	**J7**
325(322)	Conidia unicellular	*Pseudosclerophoma*	**N19**
	Conidia bicellular	*Didymochaeta*	**N20**
326(321)	Pycnidium of *textura intricata*	*Plasia*	**O2**

Pycnidium of *textura angularis*		327
327(326) Paraphyses present, filiform		328
Paraphyses absent		329
328(327) Conidia 1- to 3-septate with an extracellular, trifurcated appendage	*Pseudorobillarda*	O3
Conidia 0-septate without appendage	*Plectophomella*	O4
329(327) Conidia coloured		330
Conidia hyaline		338
330(329) Conidia surrounded by a thick, granular sheath	*Tunicago*	O5
No such character		331
331(330) Conidia with mucilaginous basal appendage	*Neottiospora*	O6
Conidia without appendage		332
332(331) Conidia 0-septate		333
Conidia 1-septate		337
333(332) Conidia fusoid-falciform or fusiform		334
Conidia not fusoid-falciform		335
334(333) Conidiophores joined on a basal pad	*Coniella*	O7
No such character	*Selenophoma*	O8
335(333) Conidia in the form of a delta	*Readeriella*	(XI B4)
No such character		336
336(335) Pycnidium not ostiolate,* forming sporodochia *in vitro*	*Tubakia*	I 16
Pycnidium ostiolate, forming pycnidia *in vitro*	*Microsphaeropsis*	O9
337(332) Conidia smooth	*Pseudodiplodia*	O10
Conidia finely verruculose	*Ascochytulina*	O11
338(329) Conidia with an apical appendage		339
Conidia without appendage		341
339(338) Appendage mucous	*Phyllosticta*	(IV D15)
Appendage cellular		340
340(339) Appendage individualized by a septum	*Mastigosporella*	(XI B6)
Appendage extending the conidium, without septum	*Ciliosporella*	P1
341(338) Conidia septate		342
Conidia non-septate		346
342(341) Conidia uniseptate		343
Conidia multiseptate		344
343(342) Ostiolar tissue uniformly coloured	*Aschochyta*	P2
Ostiolar tissue strongly pigmented	*Clypeopycnis*	P3
344(342) Pycnidium superficial	*Megaloseptoria*	P4
Pycnidium immersed		345
345(344) Conidia cylindrical or fusiform	*Stagonospora*	(V C4)
Conidia filiform	*Phlyctaeniella*	(XI B8)

Continued on p. 218

*The form of the pycnidium on stems is different from Fig. I.16, which represents a thyriopycnidium, the usual form of fungus on leaves.

Table X.A. Phialosporae, Hyphales, amerosporate, hyaline

Fig	Genus	No. spp.	Geographical distribution	Mode of life, substrates and impact	Teleomorphs	Additional references	Mol. biol.
1	Aphanocladium	4	Ubiquitous	Mycoparasite		Gams 1971	X
2	Meria	1	Europe	Parasite on Larix			
3	Sesquicillium	3	Australia, Eurasia	Saprophyte on plants	Gnomonia, Pseudonectria		
4	Harposporium	26	N. Amer., Europe	Nematophagous		Myc. Res. 101: 1377 (1997)	
5	Acremonium	105	Cosmopolitan	Telluric, saprophyte on various substrates, parasite on animals and myxomycetes, fungal antagonist, mycoherbicide, phytoparasite or endophyte, toxinogen	Calonectria, Cordyceps, Emericellopsis, Epichloe, Hapsidospora, Mycoarachis, Mycocitrus, Nectria, Neocosmospora, Niesslia, Nigrosabulum, Peckiella, Torrubiella, Trichosphaerella, Trichosphaeria, Valetoniellopsis	Gams 1971	X
6	Gliocephalis	2	N. Amer., Eurasia	1 sp. telluric and on root, 1 sp. lichenicolous			
7	Trichoderma	36	Cosmopolitan	Telluric, saprophyte on plant debris, phytoparasite and mycoparasite, fungal antagonist	Hypocrea, Podostroma, Thuemenella	CJB 69: 2357 (1992); Stud. Mycol. 41 (1998); KH 1998	X
8	Gliocladium	13	Cosmopolitan	Telluric, saprophyte on plant debris, mycoparasite, fungal antagonist	Hypocrea, Hypomyces, Nectria, Sphaerostilbella, Roumegueriella	KH 1998	X
9	Coryne	1	Ubiquitous, mostly temperate	Lignicolous, endophyte	Ascocoryne	Stud. Mycol. 31 (1989)	
10	Gloiosphaera	2	N. Amer., Europe	Lignicolous			
11	Uncigera	1	Europe	Saprophyte on litter			
12	Verticillium	40	Cosmopolitan	Telluric, phytopathogen, saprophyte, phytotoxinogen, nematophagous, entomopathogen, mycoparasite	Calonectria, Cordyceps, Ephemeroascus, Hypocrea, Hypomyces, Nectria, Torrubiella	Gams 1971; NJPP 94: 123 (1988)	X
13	Onychophora	1	Gr. Britain	Coprophilous			
14	Tolypocladium	10	Ubiquitous	Telluric, saprophyte on plants	Cordyceps		
15	Harziella	1	Europe	Fungicolous			
16	Exochalara	3	Ubiquitous	Saprophyte on wood and rotted bark or on dead stems and leaves of herbaceous plants		Fol. G. Pr. 19: 387 (1984)	
17	Sagenomella	10	America, Eurasia	Telluric, parasite or saprophyte on various substrates	Sagenoma	Pers. 10: 97 (1978)	
18	Aspergillus	185	Cosmopolitan	Telluric, saprophyte or parasite on various substrates, toxinogen	Chaetosartorya, Dichalaena, Edyuillia, Emericella, Eurotium, Fennellia, Hemicarpenteles, Leucothecium, Neosartorya, Petromyces, Sartorya, Warcupiella	Myc. Pap. 161 (1989)	X
19	Rapena	1	Indonesia	Telluric	Talaromyces		
20	Penicillium	223	Cosmopolitan	Telluric, saprophyte or parasite on various substrates, toxinogen, antagonist	Eupenicillium, Penicilliopsis, Talaromyces, Trichocoma	Pitt 1979; Stud. Mycol. 23 (1983)	X
21	Merimbla	1	Ubiquitous, temperate North	Telluric, saprophyte on plants, aerial	Talaromyces		
22	Geosmithia	8	Cosmopolitan	Telluric, saprophyte on plants	Talaromyces	CJB 57: 2021 (1979)	X
23	Riclaretia	1	Eurasia	Telluric, saprophyte on plants			
24	Paecilomyces	41	Cosmopolitan	Telluric, saprophyte on plant substrates, parasite on insects and other animals, fungicolous	Aphanoascus, Byssochlamys, Cordyceps, Dactylomyces, Melanospora, Talaromyces, Thermoascus	Stud. Mycol. 6 (1974)	X
25	Gabarnaudia	4	Europe	Telluric, saprophyte on plant substrates, coprophilous, fungicolous	Sphaeronaemella, Viennotidia		X

Phialosporae, Hyphales, amerosporate, hyaline

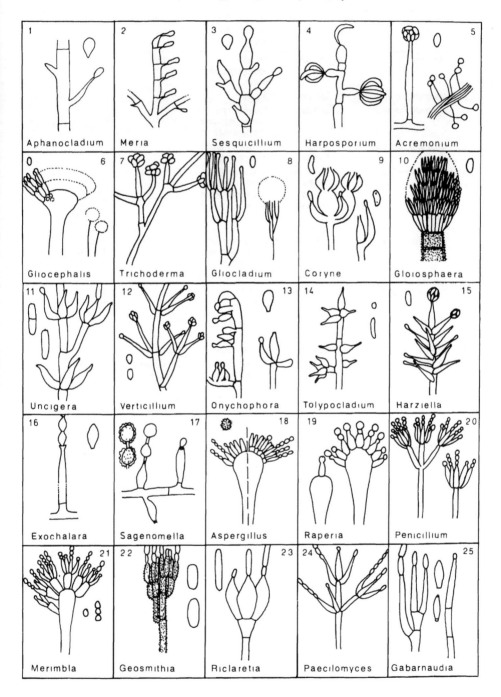

Table X.B. Phialosporae, Hyphales, didymo-phragmosporate, hyaline

Fig.	Genus	No. spp.	Geographical distribution	Mode of life, substrates and impact	Teleomorphs	Additional references	Mol. biol.
1	*Heliscus*	4	Ubiquitous	Aquatic and saprophyte on plants	*Nectria*		X
2	*Margaritispora*	5	Europe	Aquatic			
3	*Cyphellophora*	3	Europe	Parasite on human skin and phytoparasite			
4	*Cephalosporiopsis*	2	Africa, Europe	Telluric. 1 sp. nematophagous			
5	*Cylindrocladium*	23	Cosmopolitan, mostly tropical and subtropical	Phytopathogen mostly on leaves and fruits, agent of wilt	*Calonectria*	Mycot. 51: 341 (1994) S.A.M. 16: 266 (1993)	X
6	*Cylindrocladiella*	6	Cosmopolitan	Phytopathogen mostly on leaves and roots	*Nectria*	Myc. Res. 97: 433 (1993)	
7	*Cylindrodendrum*	3	Europe	Aquatic, foliicolous			
8	*Pseudomicrodochium*	9	Ubiquitous	Parasite on humans, fungicolous and saprophyte on plants		Mycot. 68: 23 (1998)	
9	*Sibirina*	6	Ubiquitous	Mycoparasite	*Hypomyces*		
10	*Fusarium*	50	Cosmopolitan	Telluric, pathogen or saprophyte on various organs of many plants, toxinogen	*Calonectria, Discostroma, Gibberella, Nectria, Nectriopsis, Plectosphaerella Sphaerulina*	GN 1983; Joffe 1986	X
11	*Cylindrocarpon*	35	Cosmopolitan	Telluric, saprophyte, facultative phytoparasite	*Nectria, Nectriopsis*		X
12	*Septofusidium*	2	Africa, Europe	Mycoparasite	*Cordyceps*		
13	*Penicillifer*	7	Ubiquitous	Telluric, radicicolous and on plant debris	*Nectria, Neocosmospora, Stellosetifera*		
14	*Cladobotryum*	9	Ubiquitous	Telluric, fungicolous	*Hypomyces*		X

Phialosporae, Hyphales, didymo-phragmosporate, hyaline

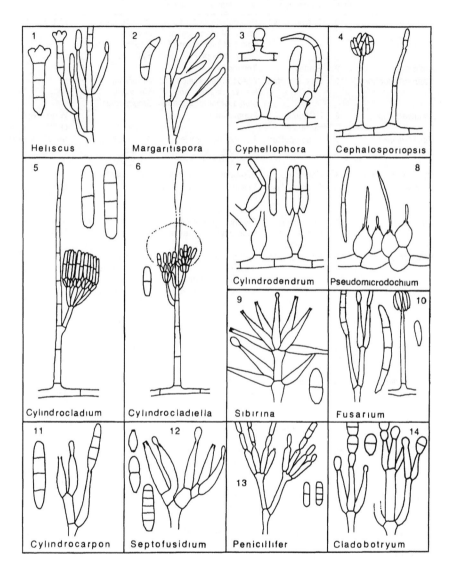

1 Heliscus	2 Margaritispora	3 Cyphellophora	4 Cephalosporiopsis

Cylindrocladium · Cylindrocladiella · Sibirina · Fusarium

Cylindrodendrum · Pseudomicrodochium

Cylindrocarpon · Septofusidium · Penicillifer · Cladobotryum

Table X.C. Melanized Hyphales, Phialosporae

Fig.	Genus	No. spp.	Geographical distribution	Mode of life, substrates and impact	Teleomorphs	Additional references	Mo bio
1	*Hormonema*	10	Cosmopolitan	Saprophyte or parasite on aerial plant organs, agent of blue stain of wood, endophyte	*Discosphaerina, Dothiora, Saccothecium, Sydowia*	AVL 65: 41 (1994)	X
2	*Anthopsis*	2	Eurasia	Telluric			
3	*Capnophialophora*	12	Cosmopolitan	Agent of fumagines on leaves, needles and stems of various plants	*Limacinia, Metacapnodium, Strigopodia*		
4	*Zakatoshia*	2	N. Amer., Europe	Fungicolous		Wind. 16: 59 (1986)	
5	*Cladorrhinum*	7	Cosmopolitan	Telluric, coprophile, saprophyte on plants, parasite on humans	*Apiosordaria, Arnium, Cercophora*	Mycot. 48: 415 (1993)	
6	*Pseudogliomastix*	1	Ubiquitous	Saprophyte on plants	*Wallrothiella*		
7	*Gliomastrix*	18	Cosmopolitan	Telluric, saprophyte on various plants, coprophile, aerial		Gams 1971	
8	*Chloridium*	17	Ubiquitous, mostly temperate	Telluric, lignicolous, saprophyte on plants	*Chaetosphaeria*	Stud. Mycol. 13 (1976)	
9	*Phialophora*	55	Cosmopolitan	Telluric, detriticolous, mostly lignicolous, parasite on plants, animals and humans, fungicolous, cases of hypovirulence	*Bombardia, Chaetosphaeria, Cistella, Coniochaeta, Dibeloniella, Gaeumanomyces, Lasiosphaeria, Mollisia, Naevala, Pyrenopeziza, Tapesia, Triangularia*		X
10	*Phaeostalagmus*	6	Ubiquitous	Telluric, saprophyte, corticolous, lignicolous		Fol. G. Pr 16: 195 (1981), Myc. Res. 96: 908 (1992)	
11	*Menispora*	10	Ubiquitous	Saprophyte on wood and bark	*Chaetosphaeria*		
12	*Dictyochaeta*	69	Cosmopolitan	Saprophyte on wood and litter	*Ascocodinaea Chaetosphaeria*	Fol. G. Pr. 19: 387 (1984); Myc. Res. 95: 1224 (1991)	
13	*Chaetopsina*	16	Cosmopolitan, mostly hot regions	Saprophyte on wood and litter	*Nectria*	TBMS 85: 709 (1985)	X
14	*Gonytrichum*	6	Ubiquitous	Telluric, saprophyte, lignicolous	*Chaetosphaeria*	Stud. Myc. 13 (1976)	
15	*Codinaeopsis*	1	Japan, USA	On wood, leaves and fruits			
16	*Custingophora*	3	Ubiquitous	Lignicolous, and on compost	*Chaetosphaeria*		
17	*Gordanichiella*	1	Ubiquitous	Telluric			
18	*Stachybotrys*	33	Cosmopolitan	Mostly saprophyte on plants, mycoparasite, toxinogen	*Melanopsamma*	Mycot. 3: 409 (1976)	X
19	*Gliocephalotrichum*	5	Ubiquitous	Telluric, saprophyte on plant debris	*Leuconectria*		
20	*Phialocephala*	21	Cosmopolitan	Mostly saprophyte on wood and litter, weak parasite on *Picea*		APS 1993; Myc. Res. 98: 745 (1994)	X

Melanized Hyphales, Phialosporae

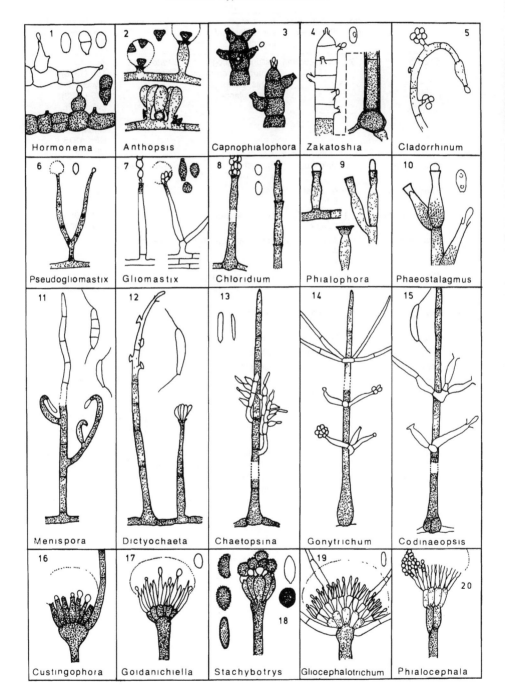

Hormonema	Anthopsis	Capnophialophora	Zakatoshia	Cladorrhinum
Pseudogliomastix	Gliomastix	Chloridium	Phialophora	Phaeostalagmus
Menispora	Dictyochaeta	Chaetopsina	Gonytrichum	Codinaeopsis
Custingophora	Goidanichiella	Stachybotrys	Gliocephalotrichum	Phialocephala

Table X.D. Melanized Hyphales, Phialosporae

Fig.	Genus	No. spp.	Geographical distribution	Mode of life, substrates and impact	Teleomorphs	Additional references	Mol. biol.
1	*Chalara*	71	Cosmopolitan	Telluric, pathogen or saprophyte on many plants, cases of hypovirulence	*Ceratocystis, Cryptendoxyla*	ADT 15: 159 (1978); Mycot. 18: 165 (1983); Fol. G. Pr. 19: 387 (1984); NRK 1975; UP 1981	X
2	*Catenularia*	6	Ubiquitous	Saprophyte on wood, bark	*Chaetosphaeria*		
3	*Torulomyces*	2	N. Amer., Europe	Telluric			
4	*Acrophialophora*	3	Australasia, Europe	Telluric, saprophyte		ABN 19: 804 (1970)	
5	*Chaetochalara*	6	Africa, N. Amer., Europe	Saprophyte on litter	*Calycellina, Hyaloscypha*		
6	*Cylindrotrichum*	12	Ubiquitous	Telluric, saprophyte on plant debris, fungicolous		Mycot. 31: 435 (1988)	
7	*Thysanophora*	5	N. Amer., Europe	Telluric, saprophyte on plant debris			
8	*Eladia*	2	Ubiquitous	Telluric			
9	*Fusariella*	10	Cosmopolitan	Telluric, lignicolous			
10	*Stachylidium*	1	Ubiquitous	Lignicolous			
11	*Sporoschismopsis*	5	Am., Eur., Australasia	Lignicolous		Myc. Res. 101: 1295 (1997)	
12	*Sporoschisma*	7	Cosmopolitan	Telluric, saprophyte on litter and wood	*Chaetosphaeria, Melanochaeta*	Myc. Res. 101: 1295 (1997)	
13	*Exophiala*	11	Cosmopolitan	Parasites on humans and animals	*Bicornispora, Capronia*		X
14	*Chaetopsis*	6	Ubiquitous	Saprophyte on litter and wood			

Melanized Hyphales, Phialosporae

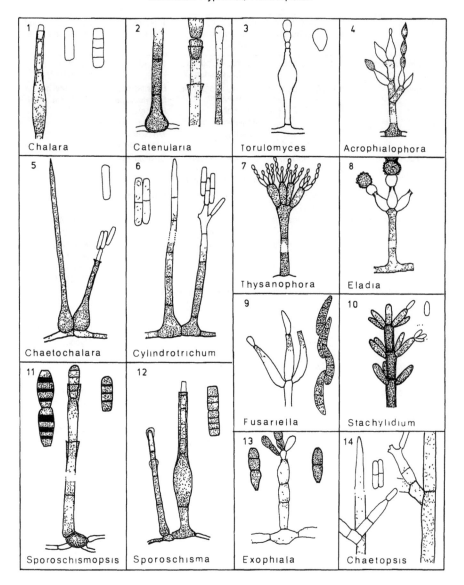

1 Chalara	2 Catenularia	3 Torulomyces	4 Acrophialophora
5 Chaetochalara	6 Cylindrotrichum	7 Thysanophora	8 Eladia
11 Sporoschismopsis	12 Sporoschisma	9 Fusariella	10 Stachylidium
		13 Exophiala	14 Chaetopsis

Table X.E. Phialosporae with coremia

Fig.	Genus	No. spp.	Geographical distribution	Mode of life, substrates and impact	Teleomorphs	Additional references	Mol. biol.
1	Dendrostilbella	15	Ubiquitous	Telluric, on plants and fungicolous	Claussenomyces	Mycol. 74: 932 (1982)	
2	Stilbella	40	Cosmopolitan	Telluric, saprophyte on plants, coprophile, fungicolous, on insects	Byssostilbe, Emericellopsis, Hypocrea, Nectria, Thyronectria, Valsonectria	Stud. Mycol. 27 (1985)	
3	Tilachlidiopsis*	3	N. and S. Amer., Europe	On rotted needles and bark, and on imago of Coleoptera	Collybia (B)		
4	Eustilbum	1	N. Amer., Europe	Resin of conifers	Bisporella		
5	Polycephalomyces	5	Cosmopolitan	On insects and Myxomycetes		Stud. Mycol. 27 (1985)	
6	Hirsutella	21	Cosmopolitan	On insects and acarids	Cordyceps, Torrubiella	Myc. Res. 94: 1111 (1990); 95: 887 (1991)	X
7	Tilachlidium	3	Cosmopolitan	On insects and fungi	Byssostilbe, Pseudonectria		
8	Didymostilbe	7	Tropical and subtropical	On plants	Peethambara	Stud. Mycol. 27 (1985)	
9	Stachycoremium	1	Japan, USA	On rotting wood		Mycol. 77: 987 (1986)	
10	Atractium	5	America, Europe	On insects			
11	Synnematium	2	Ubiquitous	On insects		TBMS 79: 431 (1982)	
12	Gibellula	8	Cosmopolitan	On arachnids	Cordyceps, Torrubiela	Mycol. 84: 300 (1992)	
13	Crinula	2	America. Europe	On living branches and on wood	Holwaya		
14	Phaeostilbella	1	Europe	On dead leaves and wood			
15	Phialographium	5	Ubiquitous	Lignicolous	Ophiostoma	APS 1993, UP 1981	
16	Saccardaea	1	America, Europe	On plants			
17	Stromatographium	1	Tropical	On dead wood	Fluviostroma	CJB 65: 2196 (1987)	
18	Pseudographiella	1	N. Amer., Europe	On rhizomorph of Armillaria			
19	Endosporostilbe	2	France, India	On dead stems and wood			
20	Chalarodendron	1	USA	On rotting wood of deciduous trees		Mycol. 76: 569 (1984)	

*Tilachlidiopisis in the present sense is considered a Phialidae (Morris, 1963). However, the study of the type species (*T. racemosa*) has shown that it is arthrosporate (CJB 69: 6, 1991). The other species are being studied (Seifert, pers. comm.).

Phialosporae with coremia

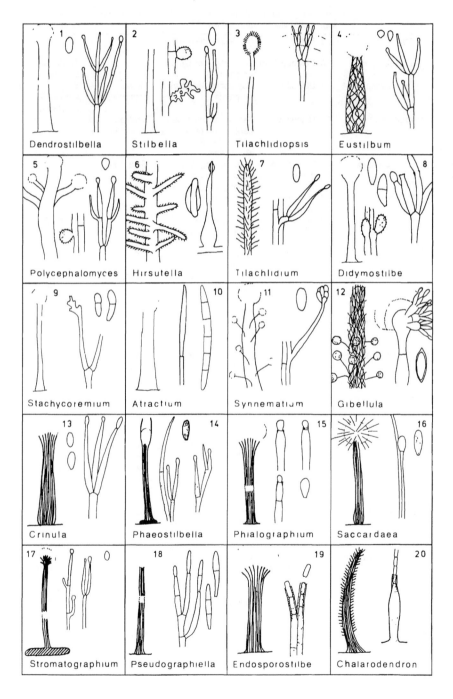

Table X.F. Phialosporae with sporodochia

Fig.	Genus	No. spp.	Geographical distribution	Mode of life, substrates and impact	Teleomorphs	Additional references	Mol. biol.
1	*Kmetia*	1	Europe	On decorticated dead wood			
2	*Sarcopodium*	6	Cosmopolitan	Saprophyte or parasite on plants		TBMS 76: 97 (1981)	
3	*Tubercularia*	25	Ubiquitous	Saprophyte or parasite on plants	*Nectria, Thyronectria*	Stud. Mycol. 27: 95 (1985)	X
4	*Volutella*	20	Ubiquitous	Telluric. saprophyte or parasite on plants, mycoparasite of *Rhizoctonia*	*Nectria, Pseudonectria*		
5	*Endoconidium*	3	Temperate zone	1 sp. pathogen of seeds of Poaceae, toxinogen	*Gloeotinia*		
6	*Endoconospora*	2	N. Europe, India	Leaf parasite			
7	*Phialophoropsis*	2	Gr. Britain, USA	1 leaf parasite. 1 symbiont of insects Scolytideae (*Ambrosia*)		TBMS 72: 337 (1979)	
8	*Tuberculina*	10	Ubiquitous	Hyperparasite of Uredinales (rust)			
9	*Myrothecium*	17	Ubiquitous	Telluric, air and plant debris, facultative parasite on plants, toxinogen	*Nectria*	Mycol. Pap. 130 (1972); Mycot. 53: 295 (1995)	X
10	*Metarhizium*	3	Cosmopolitan	Telluric, entomopathogen	*Cordyceps*	Myc. Res. 98: 225 (1994)	X
11	*Dendrodochium*	20	Mostly temperate	Telluric, saprophyte on plants	*Gnomonia, Nectria*	Myc. Pap. 130 (1972)	
12	*Actinostilbe*	2	C. and S. Amer.	Phytoparasite			
13	*Cylindrocarpon*	35	Cosmopolitan	Telluric, saprophyte or facultative phytoparasite	*Nectria, Nectriopsis*	Mycol. 85: 612 (1993)	X
14	*Fusarium*	50	Cosmopolitan	Telluric, pathogen or saprophyte on various organs of many plants, toxinogen	*Calonectria, Discostroma, Gibberella, Nectria, Nectriopsis, Plectosphaerella, Sphaerulina*	GN 1983; Joffe 1986; Mycopath. 118: 39 (1992)	X
15	*Vanbeverwijkia*	1	India, USA	Lignicolous			
16	*Cystodendron*	1	Europe	Leaf parasite maculicolous on *Quercus*	*Tapesia*		
17	*Thozetella*	6	Ubiquitous	Saprophyte or parasite on plants, foliicolous		CJB 51: 157 (1973)	
18	*Vermiculariopsiella*	6	Ubiquitous, mostly tropical	Foliicolous and on plant debris		Mycot. 37: 173 (1990)	
19	*Agyriella*	1	Europe	On branches			
20	*Bloxamia*	4	Ubiquitous	Saprophyte on plants, fungicolous	*Bisporella*	Mycot. 31: 345 (1988)	
21	*Stromatocrea*	1	N. Amer.	Corticolous and lignicolous	*Selinia*		
22	*Mycoleptodiscus*	10	Ubiquitous	Leaf parasite	*Omnidemptus*	Myc. Res. 94: 564 (1990)	
23	*Ascoconidium*	2	Temperate North	Corticolous	*Sageria*		

Phialosporae with sporodochia

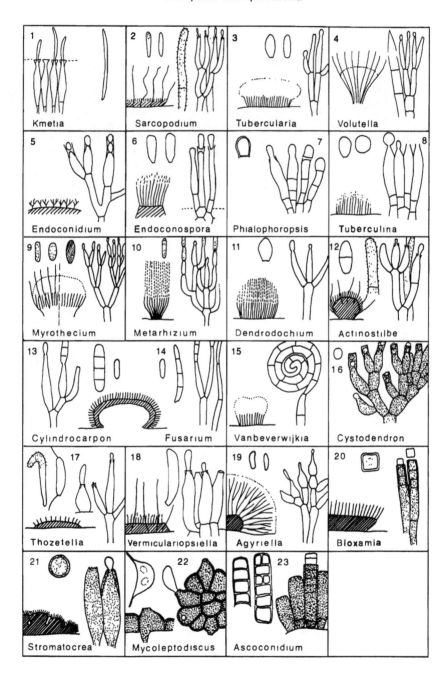

1 Kmetia	2 Sarcopodium	3 Tubercularia	4 Volutella
5 Endoconidium	6 Endoconospora	7 Phialophoropsis	8 Tuberculina
9 Myrothecium	10 Metarhizium	11 Dendrodochium	12 Actinostilbe
13 Cylindrocarpon	14 Fusarium	15 Vanbeverwijkia	16 Cystodendron
17 Thozetella	18 Vermiculariopsiella	19 Agyriella	20 Bloxamia
21 Stromatocrea	22 Mycoleptodiscus	23 Ascoconidium	

Table X.G. Acervular Phialosporae

Fig.	Genus	No. spp.	Geographical distribution	Mode of life. substrates and impact	Teleomorphs	Additional references	Mol biol.
1	*Colletotrichum*	39	Ubiquitous	Pathogen (anthracnose) on leaves. fruits. stems. seeds of many plants. agent of biocontrol on plant pests	*Glomerella*	Myc. Res. 99: 429. 475. 501 (1995)	X
2	*Polynema*	3	N. and S. Amer.. Europe	On leaves of *Asclepias* and stems of *Castanea* and *Celtis*			
3	*Greeneria*	1	Ubiquitous	Agent of bitter rot on berries of *Vitis vinifera*	*Gnomoniella*		
4	*Kaleidosporium*	1	N. Amer.	On dead shoots of *Clethra* spp.		Myc. Pap. 145 (1981)	
5	*Pycnofusarium*	1	Europe	On dead cladodes of *Ruscus aculeatus*			
6	*Titaeospora*	2	N. Amer.. Eurasia	On stems and leaves of Equisetaceae	*Stamnaria*		
7	*Septogloeum*	2	Ubiquitous	Parasite on leaves of *Celastrus* and *Euonymus*		Mycol. 85: 814 (1993)	
8	*Rhabdogloeum*	1	N. Amer.	Pathogen on needles of *Pseudotsuga*	*Rhabdocline*		
9	*Rhodesiopsis*	1	Australia, Gr. Britain	Follicolous on *Phormium*			
10	*Cytogloeum*	1	Europe	On twigs of *Tilia*			
11	*Gloeosporidiella*	11	Ubiquitous	3 spp. parasites on leaves of *Ribes* spp.. on bark of *Fraxinus*. leaves of *Laurus*...	*Drepanopeziza*		
12	*Monostichella*	9	Ubiquitous	Parasite on leaves of various plants (*Carpinus. Salix. Ficus*...)	*Drepanopeziza. Gnomonia*		
13	*Asteroma*	14	Temperate North	Maculicolous pathogen on leaves of woody plants (*Alnus, Corylus, Carpinus, Betula. Populus*)	*Gnomonia. Gnomoniella, Linospora, Pleuroceras*		
14	*Cylindrogloeum*	1	N. Amer.	Pathogen on leaves of *Trillium*			
15	*Rhodesia*	1	Germany	On stems and leaves of *Ammophila arenaria*	*Hysterostegiella*		
16	*Gloeosporidina*	6	Ubiquitous	Agent of leaf lesions on various plants including *Quercus robur*	*Apiognomonia, Stromatinia*	T.M.S. Jap. 34: 261 (1993)	
17	*Sphaceloma*	50	Ubiquitous	Agent of anthracnose and scab on various organs of many plants	*Elsinoë*		
18	*Cylindrosporium*	1	Temperate North	Leaf pathogen on *Brassica* spp.	*Pyrenopeziza*	TBMS 71: 425 (1978)	
19	*Discogloeum*	2	Europe	Parasite on leaves of *Comarum* and *Veronica*	*Spilopodia*		
20	*Discula*	12	N. Amer., Europe	Agent of anthracnose on leaves, stems and fruits of many plants (*Platanus, Quercus*...)	*Apiognomonia, Gnomonia, Gnomoniella*	Myc. Res. 96: 420 (1992)	X

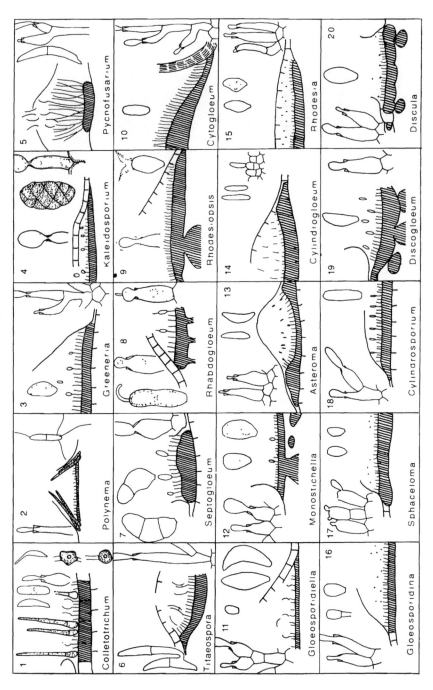

Acervular Phialosporae

1 Colletotrichum
2 Polynema
3 Greeneria
4 Kaleidosporium
5 Pycnofusarium
6 Titaeospora
7 Septogloeum
8 Rhabdogloeum
9 Rhodesiopsis
10 Cytogloeum
11 Gloeosporidiella
12 Monostichella
13 Asteroma
14 Rhodesia
15 Rhodesia
16 Gloeosporidina
17 Sphaceloma
18 Cylindrosporium
19 Cylindrogloeum
20 Discula / Discogloeum

Table X.H. Phialosporae with atypical acervuli and with cupules

Fig	Genus	No spp.	Geographical distribution	Mode of life, substrates and impact	Teleomorphs	Additional references	Mol. biol.
1	*Cheilaria*	1	Ubiquitous	Agent of leaf lesions on Poaceae			
2	*Discosporina*	4	Ubiquitous	Saprophyte on branches of *Carpinus, Carya, Corylus*			
3	*Kabatina*	4	N. Amer... Europe	Parasite on leaves and branches (3 spp. on conifers. 1 sp. on deciduous trees)		Myc. Res. 97: 1340 (1993)	
4	*Myriellina*	2	Australia. Eurasia	Parasite on living leaves of *Cydonia* and *Imperata*		Myc. Res. 95: 1021 (1991)	
5	*Pseudostegia*	1	N. Amer.	On dead leaves of *Carex*			
6	*Phacostroma*	1	Czechoslovakia	On *Ulmus*			
7	*Acleistia*	1	Europe	On female catkin of *Alnus glutinosa*	*Ombrophila*		
8	*Agyriellopsis*	2	Europe	On stripped wood of *Abies* and *Tilia*			
9	*Dinemasporium*	7	Ubiquitous	Saprophyte. foliicolous, caulicolous and corticolous	*Didymosphaeria, Keissleriella, Phomatospora*		
10	*Dwayalomella*	2	N. Amer	Parasite on stem of *Vaccinium*		Mycol. 81: 638 (1989)	
11	*Hainesia*	2	Ubiquitous	Saprophyte on leaves of many plants	*Pezizella*	Myc. Res. 101: 1228 (1997)	
12	*Hoehneliella*	1	Austria	Caulicolous on *Clematis* and *Berberis*			
13	*Pseudolachnea*	5	Ubiquitous	Saprophyte on leaves, culms and stems of many plants			
14	*Stauronema*	6	Ubiquitous	Saprophyte on leaves and stems			

Phialosporae with atypical acervuli and with cupules

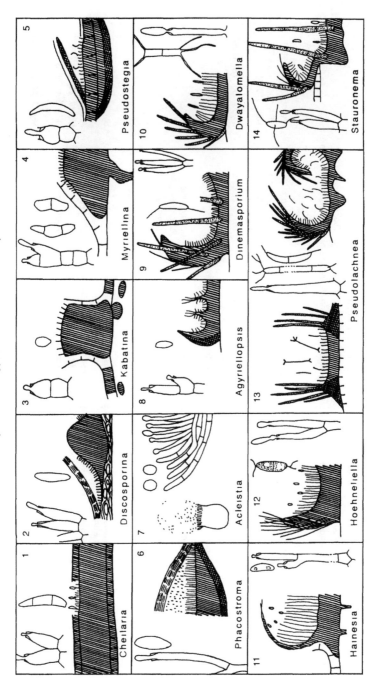

Table X.I. Pseudopycnidial Phialosporae

Fig	Genus	No. spp.	Geographical distribution	Mode of life, substrates and impact	Teleomorphs	Additional references	Mol. biol.
1	Conidioxyphum	2	Ubiquitous	On leaves of Gardenia and Oreopanax			
2	Coleophoma	4	Ubiquitous	Parasite or saprophyte on various plant organs			
3	Diachorella	4	Ubiquitous	Parasite foliicolous, caulicolous or fructicolous on Fabaceae	Diachora		
4	Melasmia	20	Ubiquitous	Parasite on living leaves (agent of tar spot on Acer spp., Salix spp., etc.)	Duplicaria, Rhytisma		
5	Myxothyrium	1	Europe	On dead leaves of Vaccinium vitis-idaea			
6	Periperidium	1	Canada	On needles of Picea mariana	Micraspis	Mycot. 21: 1 (1984)	
7	Phomopsis	100	Ubiquitous	Agent of fruit rot, dieback of stems and leaves of many plants, endophyte	Anisogramma. Diaporthe. Diaporthopsis	UFA 1988; CJB 72: 1666 (1994)	X
8	Pilidium	2	Ubiquitous	On leaves of Acer, Betula, Quercus, Rhus, Epilobium...	Discohainesia	Mycot. 83: 787 (1992)	
9	Pseudothyrium	1	Europe	On stems of Polygonatum officinale			
10	Siroplacodium	5	Eurasia	On dead stems of Campanula, Monardella, Solidago, Poaceae, Apiaceae			
11	Leptothyrium	2	Ubiquitous	On stems and siliquas of Lunaria..	Leptopeltis, Leptopeltopsis		
12	Foveostroma	6	Ubiquitous	On twigs and branches of trees	Dermea		
13	Actinothyrium	10	Ubiquitous	Saprophyte on various plant organs			
14	Rhizothyrium	2	N. and S. Amer., Europe	On leaves of Abies, Araucaria, Tsuga	Rhizocalix		
15	Tracylla	2	Ubiquitous, not in Europe	Saprophyte on leaves and twigs		Syd. 43: 264 (1991)	
16	Tubakia	5	Temperate North	Parasite maculicolous on leaves of Quercus, Castanea, Castanopsis, and on twigs of Quercus	Dicarpella	ARIF 5: 43 (1971); Mycot. 29: 101 (1987)	
17	Phlyctema	30	Ubiquitous	Saprophyte on various plant organs	Pezicula	NZJ Bot. 28: 67 (1990)	
18	Pocillopycnis	1	Sweden	On twigs of Picea abies			
19	Sirexcipula	2	N. Amer., Europe	On dead leaves of Funkia and Veratrum		Mycot. 10: 288 (1980)	

Pseudopycnidial Phialosporae

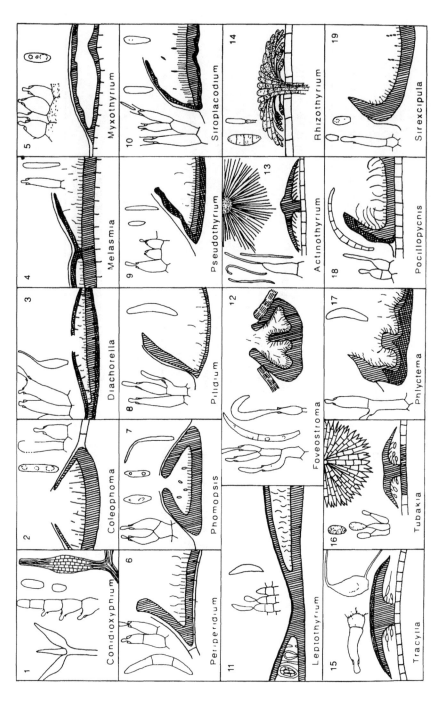

Table X.J. Pseudopycnidial Phialosporae

Fig.	Genus	No. spp.	Geographical distribution	Mode of life, substrates and impact	Teleomorphs	Additional references	Mol. biol.
1	*Amerosporium*	2	Ubiquitous	Telluric, saprophyte on various organs and debris of many plants			
2	*Botryocrea*	1	Turkey	On rotting stem of *Astragalus*		Mycol. Pap. 164 (1991)	
3	*Camarographium*	2	Europe	On petiole of *Pteridium aquilinum* and branches of *Abies* and *Picea*			
4	*Davisiella*	1	USA	Mycoparasite of *Phyllachora* on *Muhlenbergia*			
5	*Dilophospora*	1	Ubiquitous	Pathogen on leaves and culms of Poaceae	*Lidophia*		
6	*Helhonia*	1	Europe	On stem of *Sambucus nigra*			
7	*Massariothea*	8	Ubiquitous	On leaves of Poaceae and *Quercus*		Myc. Res. 97· 429 (1993)	
8	*Neoplaconema*	1	Germany, Rumania	Caulicolous on *Aconitum*			
9	*Paradiscula*	1	Europe	On stem of *Potentilla argentea*		Mycot. 2: 167 (1975)	
10	*Peltasterinostroma*	1	Gr. Britain	On dead branches of *Rubus fruticosus*			
11	*Phaeocytostroma*	4	Cosmopolitan	Telluric, parasite or saprophyte on various plant organs			
12	*Pragmopycnis*	1	Canada	On shoots of *Pseudotsuga menziesii*	*Pragmopora*		
13	*Sarcophoma*	1	Ubiquitous	On leaves of *Buxus* spp.	*Discosphaerina*	Pers. 8: 283 (1975)	
14	*Sphaeriothyrium*	2	Europe	Saprophyte on petioles of ferns			
15	*Sporonema*	15	Temperate North	Saprophyte or parasite on various plant organs	*Leptotrochila*		
16	*Zythiostroma*	2	Ubiquitous	Parasite or saprophyte on branches of conifers and *Hedera helix*	*Nectria,* *Scoleconectria*		

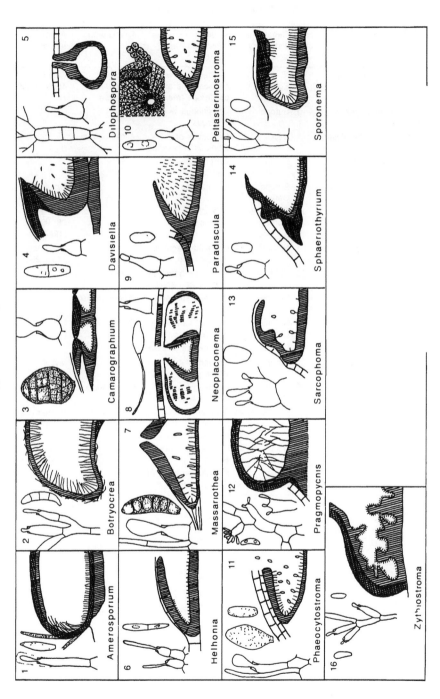

Pseudopycnidial Phialosporae

1 Amerosporium
2 Botryocrea
3 Camarographium
4 Davisiella
5 Dilophospora
6 Helhonia
7 Massariothea
8 Neoplaconema
9 Paradiscula
10 Peltasterinostroma
11 Phaeocytostroma
12 Pragmopycnis
13 Sarcophoma
14 Sphaeriothyrium
15 Sporonema
16 Zythiostroma

Table X.K. Pseudopycnidial Phialosporae

Fig.	Genus	No. spp.	Geographical distribution	Mode of life, substrates and impact	Teleomorphs	Additional references	Mol. biol.
1	*Corniculariella*	7	N. Amer., Europe	On bark and wood of deciduous and coniferous trees	*Dermea, Durandiella, Holmiella*	CJB 56: 1665 (1978)	
2	*Cornucopiella*	1	Europe	On stripped wood of *Fagus*			
3	*Cylindroxyphium*	1	USA	On leaves of *Quercus virginiana*	*Aithaloderma*		
4	*Cytonaema*	1	Europe	On branches of *Salix* spp.			
5	*Eleutheromyces*	1	Europe, Canada	Fungicolous		Crypt. Bot. 1: 384 (1990)	
6	*Sphaerographium*	10	Temperate North	Saprophyte on various plants		Myc. Res. 98: 907 (1994)	
7	*Stegonsporiopsis*	1	N. Amer.	On dead branches of *Abies* spp.		Myc. Pap. 145 (1981)	
8	*Topospora*	10	Ubiquitous	Chiefly on branches of trees and shrubs	*Godronia*		
9	*Apomelasmia*	1	Europe	On dead stems of *Urtica dioica*	*?Diaporthopsis*		
10	*Cryptomycella*	1	N. Amer., Europe	On fronds of *Pteridium* spp.	*Cryptomycina*	Mycot. 21: 1 (1984)	
11	*Gelatinopycnis*	1	Germany	On needles of *Larix*			
12	*Xyloglyphis*	1	Europe	On herbaceous stems		Mycot. 5: 87 (1977)	
13	*Cyclodomus*	2	S. Amer., USA	On dead leaves of *Umbellularia*	*Maculatifrondes*		
14	*Dothichiza*	15	Cosmopolitan	Saprophyte on branches of various trees and shrubs	*Dothiora, Saccothecium*	N. Hedw. 23: 679 (1972)	
15	*Gelatinosporium*	1	USA	On branches of *Betula*		CJB 56: 1665 (1978)	
16	*Strasseriopsis*	1	Japan	Parasite on twigs of *Tsuga sieboldi*			

Pseudopycnidial Phialosporae

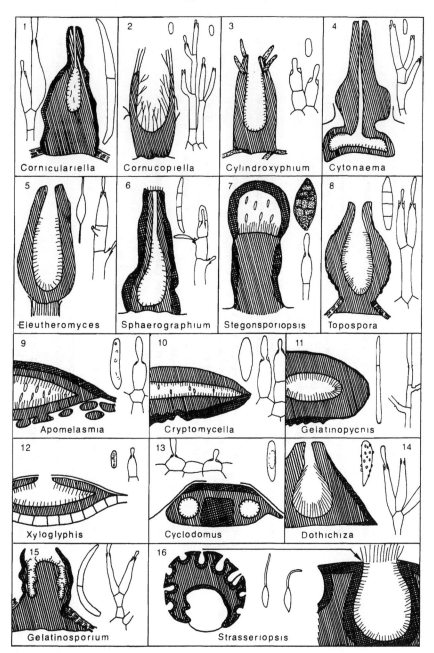

Table X.L. Pseudopycnidial Phialosporae

Fig.	Genus	No. spp.	Geographical distribution	Mode of life, substrates and impact	Teleomorphs	Additional references	Mol. biol.
1	*Allantophomopsis*	7	Ubiquitous	Parasite on leaves and branches of many conifers and some angiosperms	*Phacidium*	Mycot. 21: 1 (1984), CJB 68: 2283 (1990)	
2	*Aschersonia*	20	Tropical zone	Entomopathogen	*Hypocrella*		
3	*Asteromellopsis*	1	Switzerland	On dead stem of *Clematis vitalba*	*Dothidella*		
4	*Blennoria*	1	Europe	On dead leaves of *Buxus sempervirens*			
5	*Brunchorstia*	1	N. Amer., Eurasia	Pathogen on buds, shoots and branches of conifers	*Gremmeniella*	EJFP 10: 268 (1980)	X
6	*Camaropycnis*	1	USA	On branches of *Libocedrus* and *Pinus*			
7	*Ceuthodiplospora*	1	Czechoslovakia	On twigs of *Robinia pseudacacia*	*Pleomassaria*		
8	*Ceuthospora*	100	Cosmopolitan	Foliicolous, fructicolous, corticolous, lignicolous on many plants	*Phacidium, Pseudophacidium*	Mycot. 21: 1 (1984)	
9	*Chaetoconis*	2	Europe, USA	On stems of *Polygonum, Rumex* and *Vaccinium*	*Ceriospora*		
10	*Cytoplacosphaeria*	1	Europe	On culms of *Phragmites* spp.			
11	*Cytospora*	100	Cosmopolitan	Parasite or saprophyte on many plants, cases of hypovirulence	*Leucostoma, Valsa, Valseutypella*	GVR 1982	X
12	*Cytosporella*	30	Temperate zone	Saprophyte on many plants			
13	*Diplodina*	3	Ubiquitous	On twigs and branches of *Aesculus, Acer, Castanea* and *Salix* spp.. endophyte	*Cryptodiaporthe*		
14	*Dothiorina*	3	Europe	Lignicolous	*Chlorociboria*	Syd. 29: 146 (1977)	
15	*Endocoryneum*	2	Austria, Sardinia	On branches of *Fraxinus excelsior*, on bark of *Quercus suber*			
16	*Gyrostroma*	3	N. and S. Amer., Russia	Saprophyte on bark of *Abies, Acacia. Carya*			
17	*Microdiscula*	1	Europe	On dead stem of *Rubus fruticosus*			
18	*Nattrassia*	1	Ubiquitous	Pathogen on plants and animals		Myc. Res. 93: 483 (1989)	
19	*Placonemina*	1	Czechoslovakia	On culms of *Phragmites communis*			
20	*Pleurocytospora*	2	Europe	On branches of *Lycium* and *Ribes*			

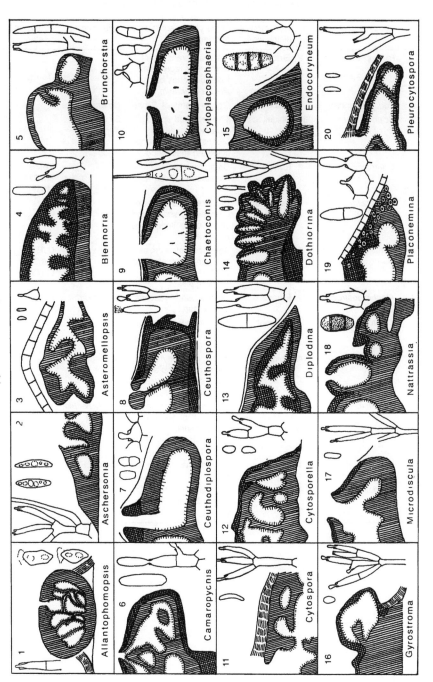

Pseudopycnidial Phialosporae

1 Allantophomopsis
2
3 Asteromellopsis
4
5 Brunchorstia
6 Camaropycnis
7 Aschersonia
8 Ceuthospora
9 Blennoria
10
11 Cytospora
12 Cytosporella
13 Diplodina
14 Chaetoconis
15 Cytoplacosphaeria
16 Gyrostroma
17 Microdiscula
18
19 Placonemina
20 Pleurocytospora
Endocoryneum
Dothiorina
Nattrassia
Ceuthodiplospora

Table X.M. Pseudopycnidial Phialosporae

Fig.	Genus	No. spp.	Geographical distribution	Mode of life, substrates and impact	Teleomorphs	Additional references	Mol. biol.
1	*Pleuroplaconema*	2	Europe, India	On dead branches of *Sambucus racemosa*, living branches of *Punica granatum*			
2	*Podoplaconema*	1	Europe	On dry stems of *Silene*	*Omphalospora*		
3	*Sclerophoma*	30	Ubiquitous	1 widespread sp. parasite on various organs of conifers and colouring their wood	*Sydowia, Xenomeris*		
4	*Scleropycnis*	2	Europe	Corticolous on branches of *Picea abies*		TBMS 88: 271 (1987)	
5	*Sirococcus*	2	Temperate zone	Parasite or saprophyte on needles, shoots and cones of conifers, twigs of *Spiraea*			
6	*Strasseria*	1	Ubiquitous	Agent of leaf necrosis and canker on stems of conifers, and of black rot of apples		CJB 57: 1660 (1979)	
7	*Amphicytostroma*	1	Europe	On moribund branches of·*Tilia* spp.	*Amphiporthe*		
8	*Cryptostroma*	1	N. Amer., Europe	Agent of sooty bark of maples (*Acer* spp.), allergen			
9	*Cytoplea*	4	Ubiquitous	3 spp. saprophytes on culms of *Arundo* and *Phyllostachys*, 1 leaf parasite of *Phragmites*	*Roussoella*		
10	*Endothiella*	2	Ubiquitous	Pathogen on branches and stem of *Castanea* and *Eugenia*, endophyte, cases of hypovirulence	*Cryphonectria, Endothia*		X
11	*Fuckelia*	1	Europe	On branches of *Ribes* spp.	*Godronia*		
12	*Sirodothis*	4	Temperate N. zone	On dead branches of *Alnus, Betula, Populus, Salix*	*Tympanis*	CJB 53: 521 (1975)	

Pseudopycnidial Phialosporae

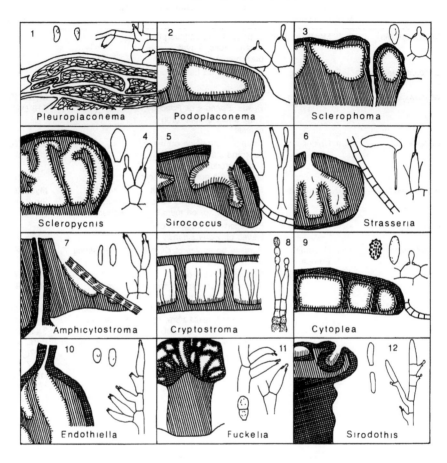

Table X.N. Pycnidial Phialosporae

Fig.	Genus	No. spp.	Geographical distribution	Mode of life. substrates and impact	Teleomorphs	Additional references	Mol. biol.
1	Chaetodiplodia	9	Mostly tropical	Saprophyte on various plant organs (fruits, leaves, branches...)			
2	Chaetosphaeronema	3	Ubiquitous	Saprophyte on herbaceous stems			
3	Wojnowicia	3	Ubiquitous	On leaves, stems or culms of *Buxus, Ephedra* or *Poaceae*, 1 sp. agent of foot-rot in cereals		Mycol. 87: 518 (1995)	
4	Chaetomella	4	Ubiquitous	Telluric, saprophyte on leaves, bark and stems of woody plants, producer of chaetomellic acid (anti-cancer)		TBMS 66: 297 (1976)	
5	Pyrenochaeta	10	Ubiquitous	Pathogen or saprophyte on leaves, stems, stocks or roots of plants, 1 sp. on toenail	*Didymella, Herpotrichia, Leptosphaeria*	Myc. Res. 99: 274 (1995)	
6	Chaetoseptoria	1	N. Amer.	Pathogen maculicolous on leaves of *Vigna sinensis*			
7	Chaetosticta	1	N. Amer., Europe	Parasite on leaves of *Cirsium* and *Artemisia*	*Nematostoma*		
8	Phoma	41	Cosmopolitan	Telluric, parasite or saprophyte on various organs of a great many plants, present on various substrates: oyster shell, glass fibre, cattle feed, droppings, cement, stone, paintings, wool...	*Cucurbitaria, Didymella, Leptosphaeria, Mycosphaerella, Pleospora, Preussia, Protocucurbitaria, Rhytidhysteron, Scutula, Tremateia, Trematosphaeria, Westerdykella*	TBMS 73: 289 (1976); Pers. 15: 197 (1993); Stud. Mycol. 3 (1973); Pers. 6; 1 (1970); Pers. 15: 71 (1992); Pers. 15: 369 (1993); Mycot. 48: 389 (1993); Pers. 17· 81 (1998)	X
9	Ampelomyces	1	Ubiquitous	Hyperparasite of Erysiphales			X
10	Amarenographium	1	Europe, USA	On leaves and culms of *Poaceae*	*Amarenomyces*	CJB 67· 3169 (1989)	
11	Ceratopycnis	1	Europe	On dead stems of *Clematis vitalba*		Mycot. 2: 167 (1975)	
12	Metazythiopsis	1	Algeria, France	Parasite on cankers of branches of *Pinus halepensis*		ASNT 40: 41 (1988)	
13	Eleutheromycella	1	Europe	Fungicolous			
14	Choanatiara	2	Canada, India	On dead needles of *Pinus* and capsules of *Eucalyptus*			
15	Pycnidiella	1	Ubiquitous	Resinicolous	*Sarea*	CJB 59: 357 (1981)	
16	Rhodosticta	3	Asia, USA	Agent of leaf lesions	*Polystigma*		
17	Phacostromella	1	Europe	On bark of branches of *Populus pyramidalis*			
18	Epithyrium	1	Ubiquitous	Resinicolous	*Sarea*	CJB 59: 357 (1981)	
19	Pseudosclerophoma	1	Czechoslovakia	On dead stems of *Acer negundo*			
20	Didymochaeta	1	N. Amer.	On dead stems of *Frasera spectosa*			

Pycnidial Phialosporae

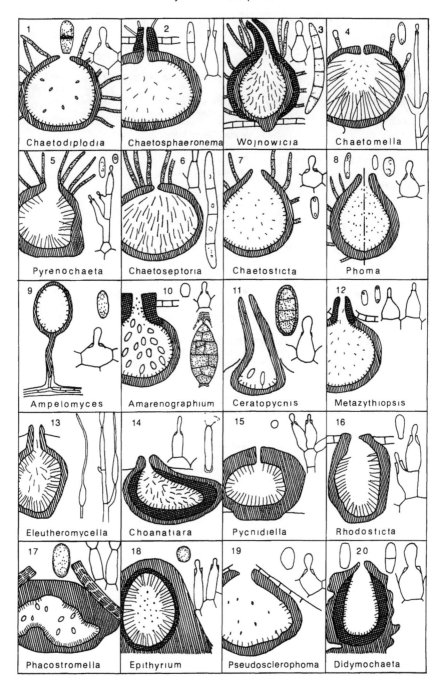

Table X.O. Pycnidial Phialosporae

Fig.	Genus	No. spp.	Geographical distribution	Mode of life. substrates and impact	Teleomorphs	Additional references	Mol. biol.
1	*Chaunopycnis*	1	Sweden	Telluric			X
2	*Plasia*	1	Europe	On branches of *Acer*	*Xylogramma*	TBMS 77: 197 (1981)	
3	*Pseudorobillarda*	9	Ubiquitous	Saprophyte on leaves, stems or glumes of Monocotyledons and Dicotyledons			
4	*Plectophomella*	1	Europe	Parasite on leaves of *Viscum album*		TBMS 77: 381 (1981)	
5	*Tunicago*	1	USA	Parasite on leaves and inflorescences of *Uniola*			
6	*Neottiospora*	1	Ubiquitous	On leaves of *Carex*, *Iris*, *Typha*			
7	*Coniella*	9	Ubiquitous	Parasite foliicolous, fructicolous on woody plants, endophyte	*Schizoparme*		
8	*Selenophoma*	8	N. Africa, Eurasia	Parasite or saprophyte on leaves and stems	*Discosphaerina*	O.U.P. 22, 24 Myc. Res. 99: 1199 (1995)	
9	*Microsphaeropsis*	44	Cosmopolitan	Present in the air, in water, soil, litter, parasite or saprophyte on various organs of many plants	*Aaosphaeria*, *Massarina*, *Paraphaeosphaeria*		
10	*Pseudodiplodia*	45	Ubiquitous	Saprophyte, caulicolous corticolous, lignicolous		Myc. Pap. 156 (1987)	
11	*Ascochytulina*	1	Europe	Saprophyte on stems of *Lonicera*			

Pycnidial Phialosporae

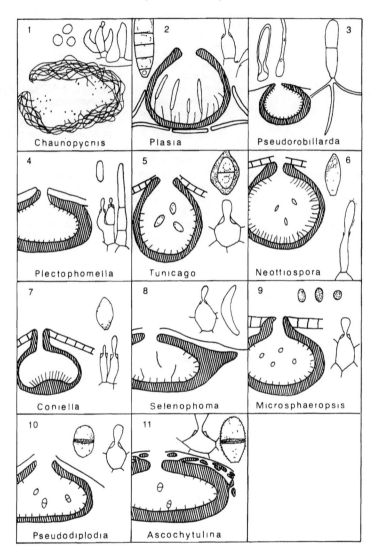

1 Chaunopycnis

2 Plasia

3 Pseudorobillarda

4 Plectophomella

5 Tunicago

6 Neottiospora

7 Coniella

8 Selenophoma

9 Microsphaeropsis

10 Pseudodiplodia

11 Ascochytulina

Table X.P. Pycnidial Phialosporae

Fig.	Genus	No. spp.	Geographical distribution	Mode of life, substrates and impact	Teleomorphs	Additional references	Mol. biol.
1	*Ciliosporella*	1	Austria	Saprophyte on dead stems of *Trifolium alpestre*			
2	*Ascochyta*	350	Cosmopolitan	Pathogen maculicolous on various organs of many plants, mycoparasite	*Didymella, Didymosphaeria, Gilletiella, Keissleriella, Mycosphaerella*	Myc. Pap. 159 (1988); Mycot. 42: 53 (1991)	X
3	*Clypeopycnis*	1	N. Amer	Saprophyte on dead stems of *Ribes*			
4	*Megaloseptoria*	1	Europe	Parasite on buds of *Picea*	*Gemmamyces*		
5	*Chaetophoma*	30	Ubiquitous	Chiefly agent of leaf fumagine			
6	*Rhizosphaera*	6	Ubiquitous	Parasite or saprophyte on needles of conifers	*Phaeocryptopus*	Rev. Myc. 43: 81 (1979); Crypto. Myc. 1: 69 (1980)	
7	*Hapalosphaeria*	1	Europe	Parasite of anthers of *Rubus*			
8	*Amerosporiopsis*	1	Iran	Saprophyte on dead leaves of *Sesleria*			
9	*Pleurophoma*	8	Ubiquitous	Lignicolous			
10	*Dendrodomus*	1	Europe	Saprophyte on dead stems of *Scrophularia*			
11	*Leptodothiorella*	20	Ubiquitous	Leaf parasite maculicolous on many plants	*Botryosphaeria, Guignardia*	Stud. Mycol. 5 (1973)	
12	*Sirophoma*	3	Europe	Lignicolous			
13	*Phialophorophoma*	1	Europe, USA	On driftwood			
14	*Asteromella*	140	Ubiquitous	Leaf parasite on many plants	*Didymosphaeria, Gillotia, Microcyclus, Mycosphaerella*	Pers. 17: 47 (1998)	
15	*Pleurophomopsis*	6	Ubiquitous	Lignicolous			
16	*Aposphaeria*	100	Ubiquitous	Lignicolous	*Melanomma, Rhytidhysteron*		

Pycnidial Phialosporae

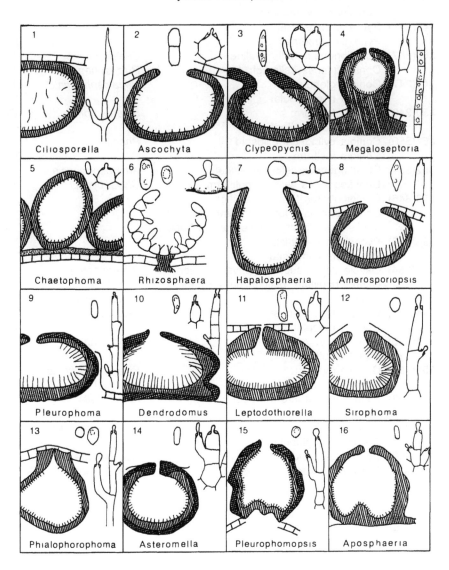

1 Ciliosporella	2 Ascochyta	3 Clypeopycnis	4 Megaloseptoria
5 Chaetophoma	6 Rhizosphaera	7 Hapalosphaeria	8 Amerosporiopsis
9 Pleurophoma	10 Dendrodomus	11 Leptodothiorella	12 Sirophoma
13 Phialophorophoma	14 Asteromella	15 Pleurophomopsis	16 Aposphaeria

346(341)	Conidiophores absent		347
	Conidiophores present		353
347(346)	Pycnidia superficial		348
	Pycnidia immersed		349
348(347)	Conidiogenous cells hyaline	*Chaetophoma*	**P5**
	Conidiogenous cells light brown	*Rhizosphaera*	**P6**
349(347)	Conidia globular	*Hapalosphaeria*	**P7**
	Conidia non-globular		350
350(349)	Conidia falciform	*Selenophoma*	**O8**
	Conidia non-falciform		351
351(350)	Conidia cylindrical ellipsoid	*Phoma*	**N8**
	Conidia fusiform		352
352(351)	Conidiogenous cells cylindrical	*Amerosporiopsis*	**P8**
	Conidiogenous cells ampulliform	*Clypeopycnis*	**P3**
353(346)	Conidiophores filiform, ramified		354
	Conidiophores filiform or cylindrical, not ramified or slightly ramified at the base		359
354(353)	Pycnidium superficial		355
	Pycnidium immersed		356
355(354)	Ostiole depressed, paraphyses absent	*Pleurophoma*	**P9**
	Ostiole papillate, ostiolar paraphyses present	*Dendrodomus*	**P10**
356(354)	Conidia acrogenous	*Leptodothiorella*	**P11**
	Conidia acropleurogenous		357
357(356)	Conidia non-guttulate	*Sirophoma*	**P12**
	Conidia guttulate		358
358(357)	Conidia oval	*Phialophorophoma*	**P13**
	Conidia fusoid	*Pyrenochaeta*	**N5**
359(353)	Pycnidium set into the substrate	*Asteromella*	**P14**
	Pycnidium superficial		360
360(359)	Conidia globular	*Pleurophomopsis*	**P15**
	Conidia non-globular		361
361(360)	Conidia ellipsoid	*Aposphaeria*	**P16**
	Conidia pyriform	*Trematophoma*	(XI B5)

ANNELLIDAE
(Enteroblastic Annellosporae)

We have mentioned earlier (in the Introduction and chapter III) that we consider the group of Annellosporae, created by modern taxonomy (Hugues, 1953), to be heterogeneous and to be close, in part, to Aleuriosporae (Annellophorae and Annelloblastosporae, chapter V) and, in part, to Phialosporae (chapter X). In the latter case, at the time of production of each conidium, the top of the neck of the conidiogenous cell becomes slightly elongated: this neck can ultimately become quite long (sometimes longer than the venter of the conidiogenous cell) and presents a finely annellated appearance (Fig. XI.1). For the fertile organ thus characterized Wang (1990) retained the name annellide.* In fact, the elongation of the neck is caused by the deposit of a series of rings, very close together and more or less visible, each corresponding to the zone of separation between conidium and conidiogenous cell by a schizogenous septum.

The successive conidia may be grouped in fragile chains (false chains) or in heads.

Most Annellidae are melanized, but Gams (1974) has described *Tympanosporium*, a hyaline mycoparasite: it simultaneously has phialides with a fixed lengh, thickened at the neck, and annellides in which the rings are nearly invisible, which increase during the production of successive conidia. It appears that the type of the sporogenous cell depends on the conditions in which it exists.

The process of conidiogenesis has been studied several times under TEM, especially in the group *Scopulariopsis-Doratomyces-Trichurus*.

Example: *Trichurus spiralis* (Hammill, 1977).

---

*This term defined and recommended at Kananaskis (Kendrick, 1971a) has been in use since the beginning of the 1970s, but it was not precisely differentiated from annellophore.

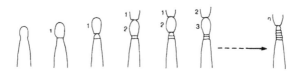

Fig. XI.1. Formation of the annellide by elongation of the neck during the production of each conidium.

In sections of annellides that have already produced a certain number of conidia, the following observations can be made (Fig. XI.2):

The conidial wall in continuity with the inner, fundamental, layer B1 of the annellide.

Successive thickenings, in a funnel shape, embedded one inside another, formed from the external layer B2.

A septation at the base of the newly formed conidium, produced a little above the last thickening of the neck of the annellide (Fig. XI.2b). At this level the wall of the conidium and of the top of the neck (constituted of the fundamental layer B1 of the annellide), upon contact with the air, becomes differentiated into internal layer B1 and external B2. This last will constitute, partly, the external layer of the wall of the conidium, and partly the new layer that will be added to the preceding layers at the top of the neck, which make up the rings.

Septations of the hyphal type with a central pore and Woronin body.

The wall of the conidium differentiates further by thickening considerably (Fig. XI.2c-d). The contents are formed principally from lipid drops in a dense cytoplasm.

In the conidial wall, at the hilum, a darkened layer more or less similar to B2 is regenerated at the exterior of B1- B′ (Fig. XI.2c-d).

A new conidium may be produced before the preceding one falls off (from which arises the formation of fragile chains). The layers B′ and B1 form the wall of the new conidium. The link between the old and new conidia seems to be reduced to the thin external film A and a septal plug (Fig. XI.2d).

It is often difficult, especially in the Conidiomales, to distinguish under a light microscope the Annellidae from other fungi that have regrowths, chiefly the Annelloblastosporae (chapters III and V), but also other more specific organisms. Thus, in *Microsphaeropsis cupressacearum* (under the name *Coniothyrium cupressacearum*), the TEM image shows that the conidiogenous cells are phialides, producing several conidia at the same site, with deposits of parietal material and thickening of the neck, but sporadic, percurrent regrowth (Reisinger et al., 1977b). In the case of *Stilbella annulata*, several conidia are produced retrogressively, that is,

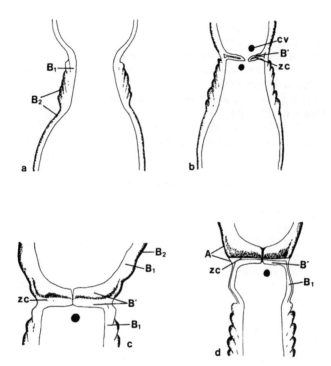

Fig. XI.2. Annellides of *Trichurus spiralis* on the basis of observations under TEM (longitudinal sections of annellides): **a:** annellide with five rings formed and production of a sixth conidium. Its wall is in continuity with the inner layer B1 of the annellide. **b:** septation between the annellide and the young conidium, with central pore and Woronin body. **c:** maturation of the conidium, with asymmetry of the walls of the conidium (thicker), and of the conidiogenous cell (thinner), on either side of the zone of cleavage. Closing of the pore. **d:** growth of the top of the conidiogenous cell to produce a new conidium, with formation of a new ring. Pore filled by a melanized substance.
A, B1, B2, B': layers of the wall. zc: zone of cleavage. cv: Woronin body.

the site of conidiogenesis goes deeper in the phialide neck before the percurrent regrowth (Roux et al., 1995).

Key to identification of Annellidae

| 1 | Hyphales | 2 |
| | Conidiomales | 19 |

Hyphales _____

| 2(1) | Fungus hyaline or subhyaline | 3 |
| | Fungus melanized | 9 |

3(2)	'Polyannellides': fertile cells with several conidiogenous openings	*Polypaecilum*	**A6**
	No such character		4
4(3)	Appearance levuriform, separate rings, conidia dividing or budding after release	*Blastoschizomyces*	**A5**
	Fungus filamentous		5
5(4)	Fertile cells long and thin = see hyphal forms of *Stilbella*		(X E2)
	Fertile cells with bulging venter		6
6(5)	Venter bulging, median between a foot and a neck	*Torulomyces*	(XD3)
	Belly basal		7
7(6)	Conidia in a barrel shape = see hyphal forms of *Tympanosporium*		**A9**
	Conidia with a pointed or pyriform tip		8
8(7)	Conidia in a mucous head	*Scedosporium*	**A1**
	Conidia in dry chains	*Scopulariopsis*	**A2**
9(2)	Conidia unicellular		10
	Conidia septate		14
10(9)	Conidia globular with germ slit	*Wardomycopsis*	**A3**
	Conidia more or less elongated without germ slit		11
11(10)	Conidiophores simple or ramified		12
	Conidiophores penicillate, long, conidiogenous cells generally sympodulate, sometimes annellate	*Leptographium*	(VI C22)
12(11)	Conidia with pointed tip	see *Spilocaea*	(V A8)
	Conidia with rounded or pyriform tip		13
13(12)	Conidia in mucous masses (cf. also *Hormonema* X C1)	*Scedosporium*	**A1**
	Conidia in dry chains	*Scopulariopsis*	**A2**
14(9)	Conidia bicellular		15
	Conidia multicellular		17
15(14)	Conidia with pointed tip	see *Spilocaea*	(V A8)
	Conidia with rounded or obtuse tips, in chain		16
16(15)	Chains and conidiophores short	*Bactrodesmiella*	(V A4)
	Chains and conidiophores long (with synanamorph *Wardomyces*: IV A30)	*Gamsia*	**A4**
17(14)	Conidiophores grouped in sporodochia	*Stigmina*	(V A7)
	Conidiphores more or less isolated on hyphae		18
18(17)	Conidiophores micronematous, dictyoconidia	*Annellophorella*	(V A3)
	Conidiophores micronematous, didymo- or phragmoconidia	*Exophiala*	(X D13)

Conidiomales

19(1)	Coremia		20
	Other conidiomata		26

Coremia

20(19)	Coremia hyaline, conidiogenous cells = annellides, sometimes sympodula	*Graphilbum*	**A11**
	Coremia melanized		21
21(20)	Conidia non-septate		22
	Conidia septate		25
22(21)	Long, non-bulging annellides		23
	Stocky annellides with bulging venter		24
23(22)	Conidiophores ramified	*Stilbella*	(X E2)
	Conidiophores non-ramified	*Graphium*	**A12**
24(22)	Fertile head of the coremium having sterile bristles	*Trichurus*	**A8**
	No such character (with or without synanamorph *Echinobotryum*: IV A29)	*Doratomyces*	**A7**
25(21)	Didymoconidia	*Bisporostilbella*	**A13**
	Phragmoconidia	*Arthrobotryum*	**A14**
26(19)	Sporodochia		27
	Other conidiomata		33

Sporodochia

27(26)	Fungus hyaline		28
	Fungus melanized		30
28(27)	Conidia unicellular, barrel-shaped	*Tympanosporium*	**A9**
	Conidia bicellular		29
29(28)	Absence of *Wardomyces* stage (aleuria with germ slit)	*Gymnodochium*	**A10**
	Presence of *Wardomyces* stage: IV A 30	*Gamsia*	**A4**
30(27)	Conidia non-septate		31
	Conidia septate		32
31(30)	Conidiophores simple, conidia pointed	*Spilocaea*	(V A8)
	Conidiophores simple or ramified, conidia rounded or pyriform	*Scopulariopsis*	**A2**
32(30)	Conidia bicellular, pointed	*Spilocaea*	(V A8)
	Conidia multicellular	*Stigmina*	(V A7)
33(26)	Acervuli		34
	Other conidiomata		39

Acervuli

34(33)	Conidia hyaline		35
	Conidia melanized		37
35(34)	Conidia bicellular	*Marssonina*	(V DI)
	Conidia unicellular		36

Continued on p. 228

Table XI.A. Annellidae, Hyphales and Conidiomales pro parte

Fig.	Genus	No. spp.	Geographical distribution	Mode of life, substrates and impact	Teleomorphs	Additional references	Mol. biol.
1	*Scedosporium*	2	Cosmopolitan	Telluric, saprophyte and parasite on humans and animals	*Pseudallescheria*	Mycot. 21: 247 (1984)	X
2	*Scopulariopsis*	31	Cosmopolitan	Telluric, saprophyte on plant debris, food, etc., pathogen on humans and animals	*Arachniotus, Chaetomium, Kernia, Microascus, Petriella, Pithoascus*		X
3	*Wardomycopsis*	3	N. Amer., Eurasia	Telluric	*Microascus*		
4	*Gamsia*	2	Europe, USA	1 sp. soil, air 1 sp. fimicolous			
5	*Blastoschizomyces*	1	USA	Human saliva			
6	*Polypaecilum*	2	S. Amer., Europe	Parasite on humans	*Dichotomomyces, Thermoascus*		
7	*Doratomyces*	8	Cosmopolitan	Telluric, saprophyte on plant substrates, fimicolous	*Pithoascus*		
8	*Trichurus*	4	Cosmopolitan	Telluric, saprophyte on plant substrates, fimicolous		Mycot. 23: 253 (1985)	
9	*Tympanosporium*	1	Europe	Mycoparasite			
10	*Gymnodochium*	1	Europe	Fimicolous			
11	*Graphilbum*	2	Japan, USA	Rotting branches, *Picea glauca*	*Ophiostoma*	UP 1981, APS 1993	
12	*Graphium*	30	Cosmopolitan	Pathogen (tracheomycesis of elm) or saprophyte on plant debris, agent of blue stain of woods dissemination by xylophages, cases of hypovirulence	*Kernia, Microascus, Ophiostoma, Petriella, Pseudallescheria*	UP 1981, APS 1993, Crypto. Myc. 14: 219 (1993)	X
13	*Bisporostilbella*	1	USA	Telluric			
14	*Arthrobotryum*	4	Eurasia, USA	Lignicolous			

Annellidae, Hyphales and Conidiomales pro parte

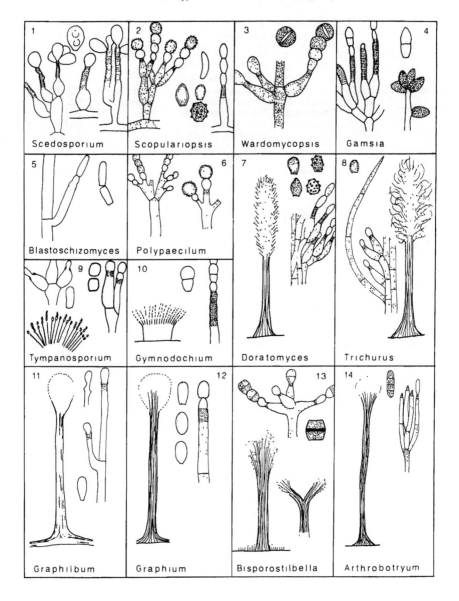

Table XI.B. Annellidae, Conidiomales

Fig.	Genus	No. spp.	Geographical distribution	Mode of life, substrates and impact	Teleomorphs	Additional references	Mol. biol.
1	*Cryptocline*	14	Temperate North	Leaf pathogen maculicolous, mostly on woody plants, endophyte	*Trochila*		
2	*Cryptosporiopsis*	17	Ubiquitous, mostly temperate	Weak parasite, agent of canker of stem, of necrosis of buds and roots, of leaf spot, endophyte, excreter of fungicidal and herbicidal substances	*Ocellaria, Pezicula*	Mycol. 85: 551, 565 (1993)	X
3	*Omega*	1	Greece	Saprophyte on dead leaves of *Quercus ilex*		TBMS 91: 715 (1988)	
4	*Readeriella*	1	Australia, Brazil, Gr. Britain	On leaves of *Eucalyptus* spp.			
5	*Trematophoma*	1	Austria	On dead wood			
6	*Mastigosporella*	2	USA	On leaves of *Quercus coccinea* and *Nyssa* spp.	*Wuestneiopsis*	Bib. Myc. 80 (1981)	
7	*Angiopomopsis*	1	Java	On dead leaves of *Phragmites*			
8	*Phlyctaeniella*	2	Europe	On dead stems of *Aruncus sylvester* and *Humulus lupulus*			
9	*Phacidiopycnis*	1	N. Amer., Europe	Agent of canker on woody *Rosaceae*	*Potebniamyces*	Mycot 21: 1 (1984)	
10	*Discosporium*	3	Ubiquitous	Parasite corticolous and foliicolous on *Populus, Salix, Eugenia*	*Cryptodiaporthe*		
11	*Cryptosporium*	25	Temperate zone	Saprophyte on many woody plants, the type species is described on *Carex* and *Poaceae*			
12	*Sphaerellopsis*	2	Ubiquitous	Hyperparasite on sori of Uredinales	*Eudarluca*		
13	*Hendersoniopsis*	1	Temperate zone	On branches of *Alnus* spp.	*Melanconis*		
14	*Neomelanconium*	2	Africa, Europe	Corticolous on trunks and branches of *Spondias* and *Tilia*			
15	*Cyclothyrium*	1	Gr. Britain, India	Saprophyte on branches of woody species	*Thyridaria*		

Annellidae, Conidiomales

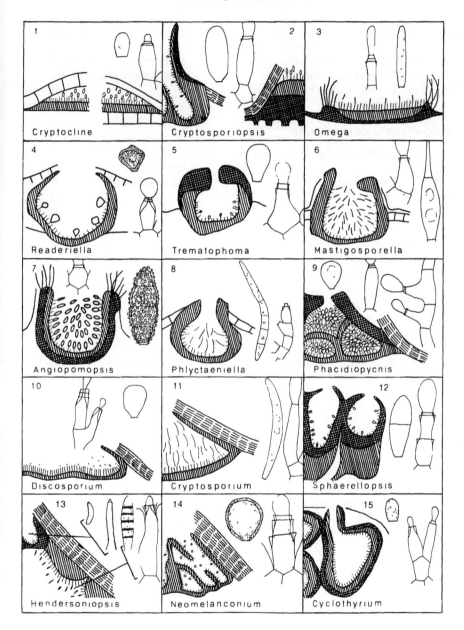

36(35)	Conidia truncated at the base	*Cryptocline*	**B1**
	Conidia abruptly retracted at the base in a well-differentiated conidial scar	*Cryptosporiopsis*	**B2**
37(34)	Conidia multicoloured brown	*Truncatella*	(V D28)
	Conidia uniformly coloured brown		38
38(37)	Conidia 0-septate, smooth	*Melanconium*	(V D11)
	Conidia 1-septate, warty	*Leptomelanconium*	(V D8)
39(33)	Conidiomata cupuloid-discoid		40
	Other conidiomata		41

Cupuloid-discoid conidiomata

40(39)	Conidiomata with brown bristles at the edge	*Omega*	**B3**
	No such character	*Pseudocenangium*	(V D25)
41(39)	Pycnidia		42
	Pseudopycnidia		51

Pycnidia

42(41)	Conidia unicellular		43
	Conidia multicellular		48
43(42)	Conidia more or less melanized		44
	Conidia hyaline		45
44(43)	Conidia deltoid with 3 apical protuberances	*Readeriella*	**B4**
	No such character	*Microsphaeropsis*	(X O9)
45(43)	Conidia without appendages	*Trematophoma*	**B5**
	Conidia with apical appendages		46
46(45)	Appendage(s) mucous	*Phyllosticta*	(IV D15)
	Appendages cellular		47
47(46)	Appendages multiple, piliform	*Neoalpakesa*	(VC6)
	Appendage single, fleshy	*Mastigosporella*	**B6**
48(42)	Conidia more or less melanized		49
	Conidia hyaline		50
49(48)	Conidia brown, without appendage	*Angiopomopsis*	**B7**
	Conidia more or less melanized, with appendages (cf. also *Hyalotiopsis* IV D13)	*Bartalinia*	(V C3)
50(48)	Conidia cylindrical, fusiform	*Stagonospora*	(V C4)
	Conidia filiform	*Phlyctaeniella*	**B8**

Pseudopycnidia

51(41)	Conidia hyaline, sometimes slightly coloured in mass		52
	Conidia melanized		58
52(51)	Conidia unicellular		53
	Conidia bicellular		56
53(52)	Conidia obovoid to ellipsoid		54
	Conidia fusiform to falciform		55

54(53)	Conidioma very stromatic, dark brown	*Phacidiopycnis*	**B9**
	Conidioma slightly stromatic, flat, light brown	*Discosporium*	**B10**
55(53)	Conidioma very stromatic, multilocular	*Fusicoccum* (IV E10), (V C13)	
	Conidioma slightly stromatic, flat, unilocular	*Cryptosporium*	**B11**
56(52)	Conidia plain	*Scaphidium*	(V C20)
	Conidia with appendages at apex		57
57(56)	Appendage long, cellular	*Uniseta*	(V C11)
	Appendage inconspicuous, mucoid	*Sphaerellopsis*	**B12**
58(51)	Conidia multicellular	*Hendersoniopsis*	**B13**
	Conidia unicellular		59
59(58)	Conidia large, globular, dark brown	*Neomelanconium*	**B14**
	Conidia small, quasi-cylindrical, light brown	*Cyclothyrium*	**B15**

XII

BASAUXIC DEUTEROMYCETES

Basauxic Deuteromycetes constitute a small group with one common characteristic: the conidiophore grows at the base (whence the term *basauxic*, from the Greek *auxein*, growth), from a specialized parent cell.

Thus, in *Dictyoarthrinium* (Cole and Samson, 1979), a parent cell appears laterally on a hypha (Fig. XII.1a, b), then a primary conidium is differentiated at its tip (Fig. XII.1c); the expansion of the conidiophore separates the primary conidium from the parent cell in a rhexolytic manner (Fig. XII.1d, e). As the conidiophore grows from the base, septa and lateral conidia appear (Fig. XII.1f-h). The septa thicken and the conidia are younger towards the base (Fig. XII.1h).

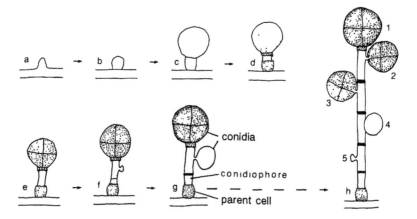

Fig. XII.1. Development of conidiophore and conidiogenesis in *Dictyoarthrinium*. Stages in the formation of the basal parent cell (**a, b**), of the primary apical conidium (**c-e**), of the intercalary conidiophore (**d-h**), and of successive lateral conidia (**f-h**).

The *Arthrinium arundinis* stage of *Apiospora montagnei* (Fig. XII.2) has been studied under electron microscope by Campbell (1975) and by Cole and Samson (1979): the primary conidium is holoblastic, as seen in Fig. XII.1c for *Dictyoarthrinium*. The conidiophore (cp) is enteroblastic, since it arises from the internal layer of the wall of the parent cell (cm) (Fig. XII.1d-e, Fig. XII.2a). This latter shows a meristematic zone at the neck. The growth continues and successive holoblastic conidia form laterally on the conidiophore. Fig. XII.2b shows a young conidium forming, having at its base a septum of the hyphal type, with a simple pore and Woronin bodies (cv), confirming that this genus belongs to the Ascomycetes. After maturity, the wall is differentiated into layers A, B and C, the basal pore is closed, and the separation occurs in a rhexolytic manner (Fig. XII.2c).

The cross-section of a conidium in Fig. XII.2c shows the presence of a hyaline ring, which delimits two valves and from which the germ tube extends. The conidia of most *Arthrinium* have this, as do those of *Cordella* and *Pteroconium*.

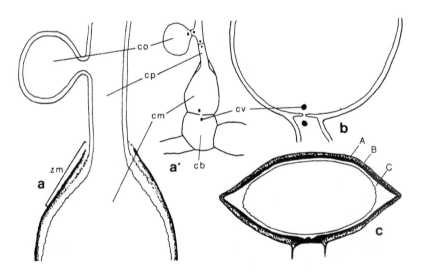

Fig. XII.2. Anamorph *Arthrinium arundinis* of *Apiospora montagnei* seen under light microscope (a') and under TEM (a, b, c) in longitudinal section.
a: growth of conidiophore from the inner layer of the parent cell wall, at the meristematic zone. Formation of a young lateral conidium.
b: partial view of a young conidium on the conidiophore, with its basal septum having a central pore and Woronin bodies.
c: mature conidium, lenticular, with melanized layers A and B and internal layer C. The hilum shows traces of a rhexolytic separation.
A, B, C: layers of the wall; **cb:** basal cell; **cm:** parent cell; **co:** conidium; **cp:** conidiophore; **cv:** Woronin bodies; **zm:** meristematic zone.]

In some species of *Arthrinium* one sees at the apex of the conidiophore one or several sterile cells, paler and smaller than the conidia and of a different shape (Table. XII.1)*

All the preceding genera have an apical conidium (or a sterile cell) and lateral conidia. They are amerosporate, with the exception of *Dictyoarthrinium,* which has dictyospores (Fig. XII.1). The sporogenous cells of the conidiophore are delimited by thick, melanized septa. In *Spegazzinia,* on the other hand, the conidiophore does not have a septum and ends in a single dictyoconidium. Moreover, often two kinds of conidia are produced, type A on long conidiophores, and type B formed on short conidiophores (Fig. XII.3).

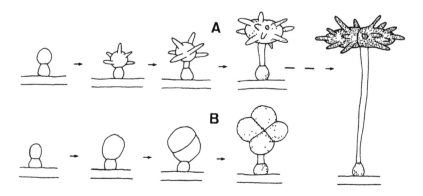

Fig. XII.3. *Spegazzinia* with two types of conidia: **A:** perpendicular to a long conidiophore; **B:** on the same plane as a short conidiophore.

Roquebert (1978) studied the formation, septation, walls, and germination of conidia in *Spegazzinia tessarthra* under TEM.

The fructifications generally affect the shape of compact or dispersed colonies and are most often superficial, except for the genera *Pteroconium, Scyphospora,* and *Endocalyx,* in which one finds respectively sporodochia, acervuli, and sessile or pedicellate cupules.

It can be remarked that it is not actually spore ontogeny that is at the basis of this group (Section VIII of Hughes, 1953)—the holoblastic conidia can be compared to aleuriospores—but the peculiarities of the parent cell and of the enteroblastic conidiophore growing from the base. Also, this parent cell has been compared to a phialide (Kendrick, 1971a; Campbell, 1975; Roquebert, 1978).

---

*Arnaud (1953) in *A. sporophloeum* and Pollack (in Tubaki, 1971) in *A. japonicum* have observed these linked terminal cells, assembled in a sort of continuous dome covering the densely serried conidiophores of a colony on a natural substrate.

Table XII. Basauxic Deuteromycetes

Fig.	Genus	No. spp.	Geographical distribution	Mode of life, substrates and impact	Teleomorphs	Additional references	Mol. biol.
1	*Arthrinium*	25	Cosmopolitan	Soil, air, plants mostly Monocotyledons, 1 sp. parasite on humans	*Apiospora* *Physalospora*		
2	*Dictyoarthrinium*	3	Tropical zone	Soil, air, plants, mostly Monocotyledons			
3	*Cordella*	4	Ubiquitous	On bamboo and carex	*Apiospora*		
4	*Pteroconium*	3	Ubiquitous	On monocotyledons	*Apiospora*		
5	*Spegazzinia*	6	Cosmopolitan	Telluric, rhizospheric, on plants, mostly Monocotyledons			
6	*Scyphospora*	1	N. Amer., Japan, New Zealand	On bamboo	*Apiospora*		
7	*Endocalyx*	4	Tropical incl. USA and Japan	On various plants		Mycol. 76: 300 (1984)	

Basauxic Deuteromycetes

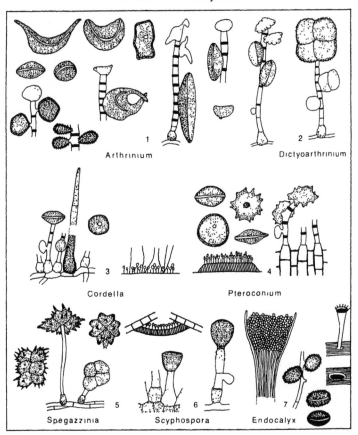

1 Arthrinium 2 Dictyoarthrinium 3 Cordella 4 Pteroconium 5 Spegazzinia 6 Scyphospora 7 Endocalyx

Close to this group are the Meristem-arthrosporae with intercalary growth of the *Erysiphe* and *Wallemia* type (Cole and Samson, 1979), which, for our part, we have discussed with the Arthrosporae (chapter II).

Key to identification of basauxic Deuteromycetes

1	Hyphales		2
	Conidiomales		5
2(1)	Amerospores		3
	Dictyospores		4
3(2)	Colonies with bristles	*Cordella*	**XII 3**
	No such character	*Arthrinium*	**XII 1**
4(2)	Conidiophores non-septate with terminal conidia	*Spegazzinia*	**XII 5**
	Conidiophores septate with terminal and lateral conidia	*Dictyoarthrinium*	**XII 2**
5(1)	Cupules pedicellate or sessile	*Endocalyx*	**XII 7**
	No such character		6
6(5)	Sporodochia	*Pteroconium*	**XII 4**
	Acervuli	*Scyphospora**	**XII 6**

---

*Scyphospora is already mentioned in chapter IV (Table IV.C18), because in nature its basauxic character does not appear to be visible. In the same way, *Polymorphum* (Table IV.E9) can be basauxic.

XIII

MYCELIA STERILIA

Sometimes also called Agonomycetes, Mycelia Sterilia do not produce conidia. They represent, like other Deuteromycetes, a stage in the life cycle of a fungus that may also have a reproductive phase with asci, basidia, or even conidia (a phase that may be known or unknown).

Among the numerous sterile mycelia that can be observed in nature or obtained in culture, for example by seeding soil or other natural substrates, some have constant characters that enable us to identify them. In particular they may form:

— A mycelium constituted of erect or monilioid hypha, isolated or more or less aggregated, in *Rhizoctonia* **sensu lato**. This last has been subdivided into several genera, belonging either to Ascomycetes (*Ascorhizoctonia* and *Complexipes*; Yang and Korf, 1985) or to Basidiomycetes (Moore, 1987). This author proposed the recognition of four genera (*Ceratorhiza, Moniliopsis, Epulorhiza* and *Rhizoctonia* s.s.) by their teleomorphs and their cytological characters. Some of these last can only be observed with the use of ultrastructural techniques or techniques of molecular biology; however, the use of aniline blue coloration can help us to observe the nuclei and the dolipores with their parenthesomes (Tu and Kimbrough, 1973). A work of identification of *Rhizoctonia* s.l. has been proposed by Sneh et al. (1991).

— Strings and clavarioid palmettes (*Anthina* Table XIII.1).

— Erect structures, more or less loose, ramified, monilioid (*Pseudaegerita*, Table XIII.7).

— Bulbils, tight groups of cells, of variable size and spherical or ovoid shape. They are produced generally in isolation, in dense masses in *Phaeosclera* (Table XIII.12), and in ramified chains in *Neoarbuscula* (Table XIII.8).

— Sclerotia, masses more organized and generally more voluminous than the bulbils, with a differentiated cortex and medulla (Fig. I.29). The cortex is melanized but the mycelium and the medulla are hyaline in *Sclerotium* (Table XIII.14). The whole thing is melanized in *Cenococcum*

(Table XIII.15). The same organization is found in the rhizomorphs, fungal structures resembling roots of small diameter, ramified, with apical meristems. These forms of conservation and propagation, often found in *Armillaria* and *Marasmius*, make up the sterile forms and were given the genus name *Rhizomorpha*, which has now fallen into disuse.

— Various propagules, ramified, complex like those of *Valdensia* (Table XIII.16) or *Cristulariella* (Table XIII.17), also considered aleuriospores by some authors because they are carried by a differentiated support that can be interpreted as a conidiophore. It is not always easy to make the distinction between these propagules and conidia; Kendrick and Carmichael (1973) classify *Mycoenterolobium* among the Mycelia Sterilia (Table IV.B26), whereas we, along with Goos (1970), the author of this genus, consider it an Aleuriosporae.

The chlamydospores are so frequent among the Fungi that it is not advisable to use this character at the generic level in the Mycelia Sterilia. The single species of the genus *Chlamydosporium*, for example, is revealed to be the form with melanized chlamydospores of *Phoma eupyrena* (Mouchacca and Sutton, 1991).

In the same way, it seems still less advisable to use the mycelium character, at the generic level, as seen in the case of *Mycelium radicis atrovirens* (sterile fungus, dark, often isolated from small roots of conifers or from the soil of boreal forests). This entity is in fact a complex of species, among which have been distinguished: *Phialocephala dimorphospora*, which fructifies very slowly (culture aged one year) and sparsely (Richard and Fortin, 1974); and, more recently, *Chloridium paucisporum*, *Phialocephala fortinii*, *Phialophora finlandia* (Wang and Wilcox, 1985), and *Leptodontidium orchidicola* (Fernando and Currah, 1995).

Key to identification of Mycelia Sterilia

1	Hyphae simple or monilioid, isolated or aggregated		2
	Bulbils, sclerotia, or other propagules		7
2(1)	Mycelium formed from erect or monilioid hyphae, isolated or assembled in small strings, sometimes sclerotioid forms	*(Rhizoctonia sensu lato)*	3
	Strings, palmettes and clavuli (in the form of small sterile Clavaria)	*Anthina*	XIII 1
3(2)	Presence of dolipores with parenthesomes, visible as bulges in the middle of hyphal septa		4
	No such character		5

4(3)	Hyphae polycaryotic, at maturity formed of barrel-shaped segments, with thick brown wall	*Moniliopsis*	**XIII 6**
	No such character	*Ceratorhiza** *Epulorhiza**	**XIII 5**
5(3)	Presence of large terminal chlamydospores on support hyphae with bulging cells	*Complexipes*	**XIII 3**
	No such character		6
6(5)	Mycelium with rapid growth, simple and aggregate hyphae of monilioid cells	*Ascorhizoctonia*	**XIII 2**
	Generally, branches at right angles, hypha retracted at the branch and septum more or less removed from it	*Rhizoctonia*	**XIII 4**
7(1)	Bulbils, sclerotia, or propagules more or less defined or in a mass		8
	Hyphae ramified at extremities in the form of acroblastospores	*Pseudaegerita*	**XIII 7**
8(7)	Papulaspores, bulbils, and sclerotia		9
	Other propagules		16
9(8)	Papulaspores: one cell or some large central cells surrounded by smaller cells	*Papulaspora*	**XIII 9**
	No such character		10
10(9)	Bulbils		11
	Sclerotia		15
11(10)	Bulbils in ramified chains, formed on a hyphal axis	*Neoarbuscula*	**XIII 8**
	Bulbils isolated or irregularly grouped		12
12(11)	Basidiomycetes with clamped hyphae, light-coloured bulbils		13
	Ascomycetes with non-clamped hyphae, brown to black bulbils		14
13(12)	Bulbil produced by multiple branches, wide and short, from a single hypha	*Burgoa*	**XIII 10**
	Bulbil formed by narrow branches of several twisted hyphae	*Minimedusa*	**XIII 11**
14(12)	Large black masses (which can reach several cm) of bulbils formed by repeated septation of an initial intercalary parent cell	*Phaeosclera*	**XIII 12**
	Bulbils resulting from repeated and dense budding of a lateral parent cell on a hypha	*Badarisama*	**XIII 13**

Continued on p. 240

---

*Ceratorhiza contains anamorphs of *Ceratobasidium*, a holobasidiomycete with perforated parenthesomes; *Epulorhiza* anamorphs of heterobasidiomycetes *Tulasnella* and *Sebacina* with non-perforated parenthesomes.

Table XIII. Mycelia Sterilia

Fig.	Genus	No. spp.	Geographical distribution	Mode of life, substrates and impact	Teleomorphs	Additional references	Mol. biol.
1	*Anthina*	5	Ubiquitous, temperate	Mostly saprophyte on litter, 2 spp. 'leaf felt' on *Citrus*	*? Cordyceps*		
2	*Ascorhizoctonia*	9	America, Eurasia	Telluric, sites of fire, saprophyte	*Tricharina*	Mycot. 24: 467 (1985). Mycot. 35: 313 (1989)	
3	*Complexipes*	3	Eurasia, USA	Forest soil, mycorrhiza	*Wilcoxina*	Mycot. 24: 467 (1985)	
4	*Rhizoctonia*	1	Ubiquitous	Agent of purple rot on roots of many plants	*Helicobasidium* (B)		X
5	*Epulorhiza* *Ceratorhiza*	3 5	Ubiquitous Ubiquitous	Root endophyte 1 sp. endophyte on root of rice, 1 sp. on aerial organs of *Eleagnus*	*Tulasnella* (B) *Sebacina* (B) *Ceratobasidium* (B)	CJB 53: 2282 (1975)	
6	*Moniliopsis*	3	Cosmopolitan	Root phytopathogen on many plants, cases of hypovirulence	*Thanatephorus* (B), *Waitea* (B)	CJB 53: 2282 (1975)	
7	*Pseudaegerita*	3	Gr. Britain, Japan, USA	Plant matter rotting in water	*Hyaloscypha*		
8	*Neoarbuscula*	1	India	On dead leaves			
9	*Papulaspora*	20	Cosmopolitan	Telluric, coprophagous, on living or decomposing plant matter, mycoparasite	*Ascobolus*, *Chaetomium*, *Melanospora*	CJB 49: 2203 (1971)	
10	*Burgoa*	4	Ubiquitous	On wood, bark, shavings, cardboard	*Cerrena* (B)	CJB 49: 2203 (1971)	
11	*Minimedusa*	1	Cuba, USA	Detriticolous (litter, old paper...), and on flower of *Gossypium*		CJB 49: 2203 (1971)	
12	*Phaeosclera*	1	Canada	Pith of *Pinus contorta*			
13	*Badarisama*	1	USA	On grains of *Soja*			
14	*Sclerotium*	100	Cosmopolitan	Telluric, saprophyte or pathogen (rotting of stems, bulbs, ...), cases of hypovirulence	*Athelia* (B), *Ciborinia*, *Sclerotinia*, *Stromatinia*	CJB 53: 2282 (1975)	X
15	*Cenococcum*	1	N. Amer., Europe	Mycorrhiza on forest trees			X
16	*Valdensia*	1	Canada, Europe	Maculicolous on leaves of *Vaccinium*, *Gaulthieria*	*Valdensinia*	CJB 50: 409 (1972)	
17	*Cristulariella*	2	Temperate North	Agent of leaf spot on mostly *Acer*	*Grovesinia*	CJB 53: 700 (1975)	

Mycelia Sterilia

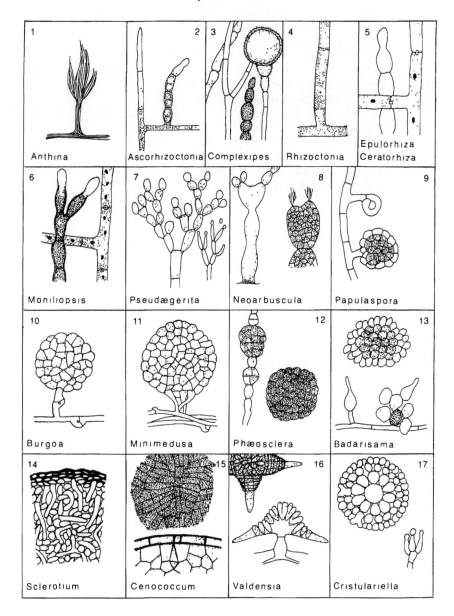

1 Anthina
2 Ascorhizoctonia
3 Complexipes
4 Rhizoctonia
5 Epulorhiza Ceratorhiza
6 Moniliopsis
7 Pseudægerita
8 Neoarbuscula
9 Papulaspora
10 Burgoa
11 Minimedusa
12 Phæosclera
13 Badarisama
14 Sclerotium
15 Cenococcum
16 Valdensia
17 Cristulariella

15(10)	Hyphae and medulla hyaline, black cortex of *textura globulosa to angularis*	*Sclerotium*	**XIII 14**
	Hyphae, cortex, and medulla black, cortex with radiated texture (elongated cells radiating from one or a few central cells)	*Cenococcum*	**XIII 15**
16(8)	Propagules in a star shape (3 to 6 arms) with active release; phialidic synanamorph *Gliocladium* (Table X.A8)	*Valdensia*	**XIII 16**
	Propagules lenticular to pyramidal, formed by repeated divisions of the apex of a propagulaphore; phialidic synanamorph *Myrioconium* and sclerotia	*Cristulariella*	**XIII 17**

BIBLIOGRAPHY

Alcorn J.L., 1983. Generic concepts in *Drechslera, Bipolaris* and *Exserohilum. Mycotaxon* **17**, 1–86.

Arnaud G., 1953. Mycologie concrète: Genera II (suite et fin). *Bull. Soc. Mycol. Fr.* **69**, 266–306.

Arx J.A., von 1970. *A revision of the fungi classified as* Gloeosporium. Bibliotheca Mycologica. **24**, J. Cramer, Lehre, 203 p.

Arx J.A., von 1981. *The genera of fungi sporulating in pure culture*, 3rd Ed. J. Cramer, Vaduz, 424 p.

Badillet G., 1991. *Dermatophyties et dermatophytes. Atlas clinique et biologique.* Ed. Varia, 303 p.

Barron G.L., 1968. *The genera of Hyphomycetes from soil.* Williams & Wilkins Co., Baltimore, 364 p.

Barron G.L., 1977. *The nematode-destroying fungi.* Topics in Mycobiology Guelph, Ont. **1**: 140 p.

Bell A., 1983. *Dung Fungi. An illustrated guide to coprophilous fungi in New Zealand.* Victoria University Press, Wellington, New Zealand, 88 p.

Booth C., (Ed.) 1971. *Methods in microbiology, Vol. 4: Methods in mycology.* Academic Press, London, New York, 795 p.

Botton B., Breton A., Fevre M., Gauthier S., Guy P., Larpent J.P., Reymond P., Sanglier J.J., Vayssier Y. Veau P., 1990. *Moisissures utiles et nuisibles, importance industrielle.* Masson, Paris, 512 p.

Brandenburger W., 1985. *Parasitische Pilze an Gefässpflanzen in Europa.* G. Fischer, Stuttgart, 1248 p.

Braun U., 1995. *A monograph of* Cercosporella, Ramularia *and allied genera (Phytopathogenic Hyphomycetes), Vol. 1.* IHW-Verlag, Eching, 333 p.

Bridge P.D., Arora D.K., Reddy C.A., Elander R.P. (eds.), 1998. *Applications of PCR in mycology.* CAB International Publishing, Wallingford, 376 p.

Brotzman H.G., Calvert O.H., Brown M.F., White J.A., 1975. Holoblastic conidiogenesis in *Helminthosporium maydis. Can. J. Bot.* 53, 813–817.

Campbell R., 1975. The ultrastructure of the *Arthrinium* state of *Apiospora montagnei.* Sacc. *Protoplasma* 83, 51–60.

Carmichael J.W., Kendrick W.B., Conners L., Sigler L., 1980. *Genera of Hyphomycetes.* Univ. Alberta Press, 386 p. (See also Seifert et al., Second Edition of this book, in Press at CBS publications).

Carroll F.E., 1971. *A fine-structural study of conidiogenesis in* Stemphylium botryosum *Wallroth = St. perf.* Pleospora herbarum *(Fr.)* Rabenh. Ph.D. Thesis, Univ. Oregon.

Carroll F.E., Carroll G.C., 1974. The fine structure of conidium initiation in *Ulocladium atrum. Can. J. Bot.* 52, 443–446.

Chabasse D., 1988. Taxinomic study of keratinolytic fungi isolated from soil and some mammals in France. *Mycopath.*, 101, 133–140.

Clauzade G., Diederich P., Roux C., 1989. Nelikenigintaj fungoj likenlogaj. Illustrita deter-
minlibro. *Bulletin de la Société Linnéenne de Provence*. No. spécial 1, 142 p.

Cole G.T., 1973. Ultrastructural aspects of conidiogenesis in *Gonatobotryum apiculatum*.
Can. J. Bot. **51**, 1677–1684.

Cole G.T., 1975. The thallic mode of conidiogenesis in the Fungi Imperfecti. *Can. J. Bot.*
53, 2983–3001.

Cole G.T., Kendrick W.B., 1968. Conidium ontogeny in Hyphomycetes. The imperfect state
of *Monascus ruber* and its meristem-arthrospores *Can. J. Bot.* **46**, 987–992.

Cole G.T., Kendrick W.B., 1969a. Conidium ontogeny in Hyphomycetes. The phialides of
Phialophora, Penicillium and *Ceratocystis. Can. J. Bot.* **47**, 779–789.

Cole G.T., Kendrick W.B., 1969b. Conidium ontogeny in Hyphomycetes. The arthrospores
of *Oidiodendron* and *Geotrichum* and the endoarthrospores of *Sporendonema. Can. J.
Bot.* **47**, 1773–1780.

Cole G.T., Samson R.A., 1979. *Patterns of development in conidial fungi.* Pitman, London,
190 p.

Coudert J., 1955. *Guide pratique de Mycologie médicale.* Masson, Paris, 364 p.

Creighton T.E. (ed.) 1999. *The encyclopedia of molecular biology.* Four volumes. Whiley
Biotechnology Encyclopedias, 3000 p.

Deighton F.C., Pirozynski K.A., 1972. Microfungi. V. More hyperparasitic hyphomycetes.
Mycological Papers No. 128, 112 p.

Dommergues Y., Mangenot F., 1970. *Ecologie microbienne du sol.* Masson, Paris, 796 p.

Domsch K.H., Gams W., Anderson T.H., 1980. *Compendium of soil fungi.* Academic Press,
London, 860 + 406 p.

Ellis D.H., Griffiths D.A., 1975a. The fine structure of conidial development in the genus
Torula. I. *T. herbarum* (Pers.) Link ex S.F. Gray and *T. herbarum* f. *quaternella* Sacc.
Can. J. Micr. **21**, 1661–1675.

Ellis D.H., Griffiths D.A., 1975b. II. *T. caligans* (Batista & Upadhyay) M.B. Ellis and *T. ter-
restris* Misra. *Can. J. Micr.* **21**, 1921–1929.

Ellis D.H., Griffiths D.A., 1977. The fine structure of conidiogenesis in *Alysidium resinae*
(= *Torula ramosa*). *Can. J. Bot.* **55**, 676–684.

Ellis M.B., 1966. Dematiaceous Hyphomycetes VII. *Curvularia, Brachysporium*, etc. *Mycol.
Pap.* **106**, 57 p.

Ellis M.B., 1971a. *Dematiaceous Hyphomycetes.* C.M.I., Kew, 608 p.

Ellis M.B. 1971b. Porospores. *In: Taxonomy of Fungi Imperfecti* (B. Kendrick, ed.) Univ. of
Toronto Press, Toronto, 71–74.

Ellis M.B., 1976. *More Dematiaceous Hyphomycetes.* C.M.I., Kew, 507 p.

Ellis M.B., Ellis J.P., 1985. *Microfungi on land plants, an identification handbook.* Croom
Helm, London, Sydney, 818 p.

Eriksson O.E., Hawksworth D.L., 1993. Outline of the Ascomycetes. *Systema Ascomycetum*
12, 51–257.

Fernando A.A., Currah R.S., 1995. *Leptodontidium orchidicola* (*Mycelium radicis atrovirens*
complex): aspects of its conidiogenesis and ecology. *Mycotaxon* **54**: 287–294.

Funk A., 1981. *Parasitic microfungi of Western trees.* Can. For Serv. Victoria, B.C., 190 p.

Funk A., 1985. *Foliar fungi of Western trees.* Can. For. Serv. Victoria, B.C., 159 p.

Galgoczy J., 1975. Dermatophytes: conidium ontogeny and classification. *Acta microbiol.,
Acad. Sci. hung.* **22**, 105–136.

Gams W., 1973. Phialides with solitary conidia? Remarks on conidium ontogeny in some
hyphomycetes. *Persoonia* **7**, 161–169.

Gams W., 1974. *Tympanosporium parasiticum* gen. et sp. nov., a hyperparasitic
Hyphomycete on *Tubercularia vulgaris* with pleomorphic conidiogenesis. *Antonie van
Leeuw.* **40**, 471–481.

Gams W., Hoekstra E.S., Aptroot A., (eds.) 1998. *CBS Course of Mycology.* Fourth edition.
CBS, Baarn, 165 p.

Goos R.D., 1970. A new genus of the Hyphomycetes from Hawaii. *Mycologia* **62**, 171–175.

Gregory P.H., 1973. *The Microbiology of the Atmosphere*, 2d ed. Leonard Hill (Books) Ltd., London, 377 p.

Griffiths D.A., 1973. The fine structure of conidial development in *Epicoccum nigrum*. *J. Microscopie* **17**, 55–64.

Griffiths D.A., 1974. Development and structure of the aleuriospores of *Humicola grisea* Traaen. *Can. J. Micr.* **20**, 55–58.

Grove W.B., 1919. Mycological notes. IV. *J. Bot. Lond.* **57**, 206–210.

Grove W.B., 1935/1937. *British stem and leaf fungi (Coelomycetes)*, Vol. 1: *Sphaeropsidales, Sphaeriodeae with hyaline conidia*, 488 p. Vol. 2: *Sphaeropsidales rest. Melanconiales*, 407 p. Cambridge, Repr. J. Cramer 1967.

Hammill T.M., 1971. Fine structure of annellophores. I. *Scopulariopsis brevicaulis* and *S. koningii. Amer. J. Bot.* **58**, 88–97.

Hammill T.M., 1972a. Fine structure of annellophores. II. *Doratomyces nanus. Trans. Brit. mycol. Soc.* **59**, 249–253.

Hammill T.M., 1972b. Fine structure of annellophores. III. *Monotosporella sphaerocephala. Can. J. Bot.* **50**, 581–585.

Hammill T.M., 1974a. Fine-structural aspects of reproduction in Fungi Imperfecti. *VIII Int. Congr. Electr. Microsc.*, Canberra II, 560–561.

Hammill T.M., 1974b. Electron microscopy of phialides and conidiogenesis in *Trichoderma saturnisporum. Amer. J. Bot.* **61**, 15–24.

Hammill T.M., 1977. Transmission electron microscopy of annellides and conidiogenesis in the synnematal hyphomycete *Trichurus spiralis. Can. J. Bot.* **55**, 233–244.

Hanlin R.T., 1976. Phialide and conidium development in *Aspergillus clavatus. Amer. J. Bot.* **63**, 144–155.

Hanlin R.T., 1982. Conidiogenesis in *Spiniger meineckellus. Mycologia* **74**, 236–241.

Hashmi M.H., Morgan Jones G., 1973. Conidium ontogeny in Hyphomycetes. The meristem-arthrospores of *Wallemia sebi. Can. J. Bot.* **51**, 1669–1671.

Hashmi M.H., Morgan Jones G., Kendrick B., 1973. Conidium ontogeny in Hyphomycetes. The blastoconidia of *Cladosporium herbarum* and *Torula herbarum. Can. J. Bot.* **51**, 1089–1091.

Hawes C.R., 1979. Conidium ultrastructure and wall architecture in the *Chalara* state of *Ceratocystis adiposa. Trans. Brit. mycol. Soc.* **72**, 177–187.

Hawes C.R., Beckett A., 1977a. Conidium ontogeny in the *Chalara* state of *Ceratocystis adiposa*. I. Light microscopy. *Trans. Brit. mycol. Soc.* **68**, 259–265.

Hawes C.R., Beckett A., 1977b. Conidium ontogeny in the *Chalara* state of *Ceratocystis adiposa*. II. Electron microscopy. *Trans. Brit. mycol. Soc.* **68**, 267–276.

Hawes C.R., Beckett A., 1977c. Conidium ontogeny in *Thielaviopsis basicola. Trans. Brit. mycol. Soc.* **68**, 304–307.

Hawksworth D.L., 1981. The lichenicolous Coelomycetes. *Bulletin of the British Museum Botany Series*, **9**, 1–98.

Hawksworth D.L., Kirk P.M., Sutton B.C., Pegler D.N., 1995. *Ainsworth & Bisby's Dictionary of the fungi*, 8th Ed. CAB International, Wallingford, 616 p.

Hennebert G.L., 1973. *Botrytis and Botrytis-like genera. Persoonia* **7**, 183–204.

Hennebert G.L., 1977. *Démonstrations mycologiques*. Louvain-la-Neuve, Ed. 97 p.

Hennebert G.L., Sutton B.C., 1994. Unitary parameters in conidiogenesis. *In: Ascomycete systematics: Problems and perspectives in the nineties*, (Hawksworth Edit.). Plenum Press, New York, 65–76.

Hennebert G.L., Weresub L.K., 1977. Terms for states and forms of Fungi, their names and types. *Mycotaxon* **6**, 207–211.

Hoog G.S., de 1974. The genera *Blastobotrys, Sporothrix, Calcarisporium* and *Calcarisporiella* gen. nov. *C.B.S. Studies in Mycology* **7**, 1–84.

Hoog G.S., de 1983. The taxonomic structure of *Exophiala. In: Fungi pathogenic for humans and animals* (D.H. Howard, ed.), Part B II, 327–336.

Hoog G.S., de 1985. *Dactylaria, Neta, Subulispora* and *Scolecobasidium* IV. *In: Taxonomy of the Dactylaria complex. C.B.S. Stud. Mycol.* **26**, 1–60.

Hogg G.S., de, Guarro J. (eds.), 1996. *Atlas of clinical fungi.* Baarn, Netherlands & Reus, Spain, 720 p.

Hugues G.C., Bisalputra A.A., 1970. Ultrastructure of Hyphomycetes. Conidium ontogeny in *Peziza ostracoderma. Can. J. Bot.* **48**, 361–366.

Hugues S.J., 1953. Conidiophores, conidia and classification. *Can. J. Bot.* **31**, 577–659.

Ilyaletdinov A.N., Pryadko E.I. (eds.) 1990. *Griby-Gifomitsety Regulyatory chislennosti Paraziticheskikh Nematod.* Alma-Ata, USSR: Nauka Kazakhskoi, 176 p.

Ingold C.T., 1975. *An illustrated guide to aquatic and water-borne hyphomycetes (Fungi Imperfecti) with notes on their biology.* Scient. Publ. Freshw. Biol. Assoc. 30, 96 p.

Ingold C.T., 1981. The first-formed phialoconidium of *Thielaviopsis basicola. Trans. Brit. mycol. Soc.* **76**, 517–519.

Innis M.A., Gelfand D.H., Sninsky J.J., White T.J. (eds.) 1990. *PCR protocols: A guide to the methods and applications.* Academic Press, New York, 482 p.

Jesenskà Z. (ed.), 1993. *Micromycetes in foodstuffs and feedstuffs.* Progress in industrial microbiology, Vol. 28 Elsevier, Amsterdam, 256 p.

Joly P., 1965. Clé de détermination des espèces les plus courantes du genre *Alternaria* (Nees.) Wiltsh. emend. Joly. *Rev. Mycol.* **29**, 348–351.

Jones J.P., 1976. Ultrastructure of conidium ontogeny in *Phoma pomorum, Microsphaeropsis olivaceum,* and *Coniothyrium fuckelii. Can. J. Bot.* **54**, 831–851.

Kendrick W.B. (ed.), 1971a. *Taxonomy of Fungi Imperfecti.* Univ. Toronto Press, Toronto, 309 p.

Kendrick W.B., 1971b. Arthroconidia and meristem-arthroconidia. *In: Taxonomy of Fungi Imperfecti* (B. Kendrick, ed.). Univ. Toronto Press, Toronto, 160–175.

Kendrick W.B. (ed.), 1979. *The Whole Fungus. The Sexual-Asexual Synthesis.* Natl. Mus. Can. Ottawa, 2 vols, 793 p.

Kendrick W.B., Carmichael J.W., 1973. Hyphomycetes. *In: The Fungi,* Vol. IV A. (G.C. Ainsworth, F.K. Sparrow, A.S. Sussmann, Eds.). Academic Press, N.Y., 323–509.

Kendrick W.B., Cole G.T., 1968. Conidium ontogeny in Hyphomycetes. The sympodulae of *Beauveria* and *Curvularia. Can. J. Bot.* **46**, 1297–1301.

Kendrick W.B., Cole G.T., 1969. Conidium ontogeny in Hyphomycetes. *Trichothecium roseum* and its meristem-arthrospores. *Can. J. Bot.* **47**, 345–350.

Kendrick W.B., Cole G.T., Bhatt G.C., 1968. Conidium ontogeny in Hyphomycetes. *Gonatobotryum apiculatum* and its botryose blastospores. *Can. J. Bot.* **46**, 591–596.

Kendrick W.B., Nag Raj T.R., 1979. Morphological terms in Fungi Imperfecti. *In: The Whole Fungus,* Vol. 1. (W.B. Kendrick, Ed.). Natl. Mus. Can. Ottawa, 43–61.

Khan S.R., Kimbrough J.W., 1982. A reevaluation of the Basidiomycetes based upon septal and basidial structures. *Mycotaxon* **15**, 103–120.

Kiffer E., Mangenot F., Reisinger O., 1971. Morphologie ultrastructurale et critères taxinomiques chez les Deutéromycètes. IV. *Doratomyces purpureofuscus* (Fres.) Morton et Smith. *Rev. Ecol. Biol. Sol.* **8**, 397–407.

Kirk P.M., 1985. New or interesting microfungi. XIV. Dematiaceous Hyphomycetes from Mt Kenya. *Mycotaxon* **23**, 305–352.

Korf R.P., 1952. A monograph of the Arachnopezizae. *Lloydia.* **14**, 129–180.

Korf R.P., Hennebert G.L., 1993. A disastrous decision to suppress the terms anamorph and teleomorph. *Mycotaxon* **48**, 539–542.

Langeron M., Vanbreuseghem R., 1952. *Précis de Mycologie.* Masson, Paris, 703 p.

Loncle D., Amaudric M., Jacoty C., 1993. *Génie génétique.* Coll. Bio-sciences et techniques. Doin éditeurs, Paris, 439 p.

Luttrell E.S., 1963. Taxonomic criteria in *Helminthosporium. Mycologia* **55**, 643–674.

Luttrell E.S., 1964. Systematics of *Helminthosporium* and related genera. *Mycologia* **56**, 119–132.

Luttrell E.S., 1969. *Curvularia coicis* and the *nodulosa* group of *Bipolaris. Mycologia* **61**, 1031–1040.

Malloch D., 1981. *Moulds, their isolation, cultivation and identification.* Univ. Toronto Press, Toronto, 97 p.

Malone J.P., Muskett A.E., 1997. *Seed-borne fungi.* Descriptions of 77 Fungus Species, 3rd ed. International Seed Testing Association, Zurich, 191 p.

Mangenot F., 1952. Recherches méthodiques sur les champignons de certains bois en décomposition. *Revue gén. Bot.,* **59**, 1–115.

Mangenot F., Reisinger O., 1976. Form and function of conidia as related to their development. *In: The Fungal Spore* (D.J. Weber, W.M. Hess, eds). John Wiley, N.Y., 789–847.

Martinez A.T., Guarro J., Figueras M.J., Punsola L., 1986. Arthric conidiogenesis in *Malbranchea arcuata. Trans. Brit. mycol. Soc.* **86**, 490–494.

Mason E.W., 1933. *Annotated account of fungi received at the Imperial Mycological Institute.* List II, Fasc. 2, I.M.I., Kew, 1–67.

Matsushima T., 1975. *Icones Microfungorum a Matsushima lectorum.* Kobe, Japan, 209 p.

Melnik V.A., 1997. *Definitorium Fungorum Rossiae. Classis Coelomycetes. Fasc. 1. Genera rara et minus cognita.* Petropoli "Nauka", 279 p.

Messiaen C.M., Blancard D., Rouxel E., Lafon R., 1991. *Les maladies des plantes maraîchères, 3rd ed.* INRA Editions, Paris, 552 p.

Minter D.W., Kirk P.M., Sutton B.C., 1982. Holoblastic phialides. *Trans. Brit. mycol. Soc.* **79**, 75–93.

Minter D.W., Kirk P.M., Sutton B.C., 1983a. Thallic phialides. *Trans. Brit. mycol. Soc.* **80**, 39–66.

Minter D.W., Sutton B.C., Brady B.L., 1983b. What are phialides anyway? *Trans. Brit. mycol. Soc.* **81**, 109–120.

Moore R.T., 1977. Dolipore disjunction in *Antromycopsis broussonetiae. Pat. Experim. Mycol.* **1**, 92–101.

Moore R.T., 1987. The genera of *Rhizoctonia*-like fungi: *Ascorhizoctonia, Ceratorhiza* gen. nov., *Epulorhiza* gen. nov., *Moniliopsis,* and *Rhizoctonia. Mycotaxon,* **29**, 91–99.

Moreau C., 1974. *Moisissures toxiques dans l'alimentation.* Masson, Paris, 471 p.

Morelet M., 1978. Deutéromycètes (Fungi Imperfecti). *In: Mycologie et Pathologie forestières. I. Mycologie forestière* (Lanier L., Joly P., Bondoux P., Bellemere A.). Masson, Paris, 367–407.

Morris E.F., 1963. *The synnematous genera of the Fungi Imperfecti.* Western Illinois University Series Biol. Sciences no. 3, 143 p.

Mouchacca J., Sutton B.C., 1991. The genus *Chlamydosporium* Peyronel. *Cryptog. Mycol.* **12**, 251–255.

Muller E., 1981. Relations between Conidial Anamorphs and their Teleomorphs. *In: Biology of Conidial Fungi,* Vol. I. (G.T. Cole, B.W. Kendrick, eds). Academic Press, N.Y., 145–169.

Nag Raj T.R., 1981. Coelomycete Systematics. *In:* Biology of *conidial fungi,* Vol. I (G.T. Cole, B.W. Kendrick, eds). Academic Press, N.Y., 43–84.

Nag Raj T.R., 1993. *Coelomycetous anamorphs with appendage-bearing conidia.* Mycologue Publications, Waterloo, Ontario, 1101 p.

Neergaard P., 1978. *Seed pathology.* MacMillan, London, Basingstoke, 1187 p.

Nicot J., 1966. Clé pour la détermination des espèces banales de champignons fongicoles. *Revue Mycol.,* **31**, 393–399.

Nilsson S., (Ed.). 1983. *Atlas of airborne fungal spores in Europe.* Springer-Verlag, 139 p.

Okada G., Tubaki K., 1986. Conidiomatal structures of the Stilbellaceous and allied fungi. *Sydowia* **39**, 148–159 (publ. 1987).

Percebois G., 1973. Introduction a une étude des dermatophytes. *Bull. Ass. Dipl. Microbiol. Fac. Pharm. Nancy,* **129**, 60 p.

Pitt J.I., 1974. A synoptic key to the genus *Eupenicillium* and to sclerotigenic *Penicillium* species. *Can. J. Bot.* **52**, 2231–2236.

Pitt J.I., Hocking A.D., 1997. *Fungi and food spoilage, 2nd ed.* Blackie Academic & Professional, London, 593 p.

Punithalingam E., 1969/74/81. Studies on Sphaeropsidales in culture, I, II, and III. *Mycol. Pap.* **119**, 1–24: **136**, 1–63; **149**, 1–42.

Raper K.B., Thom C., 1949. *A manual of the Penicillia.* Williams & Wilkins Co., Baltimore, 875 p.

Rapilly F., 1968. Les techniques de mycologie en pathologie végétale. *Annales des Epiphyties* **19**, 3–102.

Raynal G., Gondran J., Bournoville R., Courtillot M., 1989. *Ennemis et maladies des prairies.* INRA Editions, Paris, 249 p.

Reisinger O., 1966. Sur *Conoplea mangenotii* sp. nov. isolé à partir de branches mortes de *Rhus cotinus. Rev. Mycol.* **31**, 329–340.

Reisinger O., 1970a. Morphologie ultrastructurale et critères taxinomiques chez les Deutéromycètes. III./Etude au microscope électronique (à balayage et à transmission) de *Wardomyces pulvinata* (Marchal) Dickinson. *Acta. phytopat.* **5**, 221–230.

Reisinger O., 1970b. Etude aux microscopes électroniques à balayage et à transmission, de la paroi sporale et de son rôle dans la dispersion des conidies chez *Helminthosporium spiciferum* (Bain.) Nicot. *C.R. Acad. Sc. Paris, D* 270, 3031–3032.

Reisinger O., 1972. *Contribution à l'étude ultrastructurale de l'appareil sporifère chez quelques Hyphomycètes à paroi mélanisée. Genèse, modification et décomposition.* Thèse Doct. Sci. Nat. Univ. Nancy I, 360 p. miméogr.

Reisinger O., Kiffer E., Mangenot F., Olah G.M., 1977a. Ultrastructure, cytochimie et microdissection de la paroi des hyphes et des propagules exogènes des Ascomycètes et des Basidiomycètes. *Rev. Mycol.* **41**, 91–117.

Reisinger O., Morelet M., Kiffer E., 1977b. Electron microscopic study of conidium ontogeny in *Coniothyrium cupressacearum* (Coelomycetes). *Persoonia* 9, 257–264.

Richard C., Fortin J.A., 1974. Distribution géographique, écologie, physiologie, pathogénicité et sporulation du *Mycelium radicis atrovirens. Phytoprotection* **55**, 67–88.

Richardson M.J., Watling R., 1997. *Keys to fungi on dung,* rev. ed. British Mycological Society, Stourbridge, West Midlands, 68 p.

Roquebert M.F., 1978. Conidiogenèse et conidies chez *Spegazzinia tessarthra* (Berk. et Curt.) Sacc. *Rev. Mycol.* **42**, 309–320.

Roquebert M.F., 1981. Analyse des phénomènes pariétaux au cours de la conidiogenèse chez quelques champignons microscopiques. *Mém. Mus. Nat. Hist. Nat., Nouv. Ser. B.* **28**, 1–79, + 23 pl.

Roquebert M.F., Bury E., 1988. *Leptoxyphium:* pycnide ou synnema? *Can. J. Bot.* **66**, 2265–2272.

Roux C., Der Merwe C.F., van, Warmelo K.T., van 1995. The ultrastructure of conidiogenesis in *Stilbella annulata. S. Afr. J. Bot.* **61**, 215–221.

Saccardo P.A., 1880. Conspectus generum fungorum italiae inferiorum nempe ad Sphaeropsideas, Melanconieas et Hyphomyceteas pertinentium, systemate sporologico dispositoru. *Michelia Commentorium Mycologicum* **2**, 1–38.

Saccardo P.A., 1884. *Sylloge fungorum omnium hucusque cognitorum,* Vol. 3. Pavia, 860 p.

Samson R.A., Evans H.C., Latge J.P., 1988. *Atlas of entomopathogenic fungi.* Springer-Verlag & Utrecht, Berlin, 187 p.

Samuels G.J., McKenzie E.H.C., Buchanan D.E., 1981. Ascomycetes of New Zealand. 3. Two new species of *Apiospora* and their *Arthrinium* anamorphs on bamboo. *New Zeal. Jour. Bot.* **19**, 137–149.

Segretain G., Drouhet E., Mariat F., 1979. *Diagnostic de laboratoire en mycologie médicale.* Maloine S.A., Paris, 150 p.

Shoemaker R.A., 1959. Nomenclature of *Drechslera* and *Bipolaris* grass parasites segregated from "*Helminthosporium*". *Can. J. Bot.* **37**, 879–887.

Sigler L., Carmichael J.W., 1976. Taxonomy of *Malbranchea* and some other Hyphomycetes with arthroconidia. *Mycotaxon* 4, 349–388.

Simmons E.G., 1967. Typification of *Alternaria, Stemphylium* and *Ulocladium. Mycologia* **59**, 67–92.

Simmons E.G., 1969. Perfect states of *Stemphylium. Mycologia* **61**, 1–26.

Simmons E.G., 1971. *Helminthosporium allii* as type of a new genus. *Mycologia* **63**, 380–386.

Simmons E.G., 1983. An aggregation of *Embellisia* species. Mycotaxon **17**, 216–241.
Singleton L.L., Mihail J.D., Rush Ch.M. (eds.), 1993. *Methods for research on soil-borne phytopathogenic fungi*. APS Press, St. Paul, Minnesota, 265 p.
Sivanesan A., 1984. *The bitunicate Ascomycetes and their anamorphs*. J. Cramer, Vaduz, 701 p.
Sivanesan A., 1985. New species of *Bipolaris*. *Trans. Brit. mycol. Soc.* **84**, 403–421.
Smith I.M., Dunez J., Lelliott R.A., Phillips D.H., Archer S.A., 1988. *European handbook of plant diseases*. Blackwell Scientific Publications, Oxford, 583 p.
Sneh B., Burpee L., Ogoshi A., 1991. *Identification of* Rhizoctonia *species*. APS Press, St. Paul, Minnesota, 133 p.
Starback K., 1895. Discomyceten-Studien, Kgl. Sv. Vetenskaps—*Akad. Handl, Bihang.* **21**, 1–42.
Stiers D.L., Rogers J.D., Russel D.W., 1973. Conidial state of *Poronia punctata*. *Can. J. Bot.* **51**, 481–484.
Subramanian C.V., 1962. A classification of the Hyphomycetes. *Curr. Sci.* **31**, 409–411.
Subramanian C.V., 1971. *Hyphomycetes*. Indian Counc. Agric. Res., New Delhi, 930 p.
Subramanian C.V., 1983. *Hyphomycetes, taxonomy and biology*. Academic Press, London. 502 p.
Sutton B.C., 1973a. Ch. 11, Coelomycetes. *In: The Fungi*, IV A (G.C. Ainsworth, F.K. Sparrow, A.S. Sussman, eds). Academic Press, N.Y., 513–582.
Sutton B.C., 1973b. Hyphomycetes from Manitoba and Saskatchewan, Canada. *Mycol. Pap.* **132**, 143 p.
Sutton B.C., 1980. *The Coelomycetes*. C.M.I., Kew, 696 p.
Sutton B.C., 1993. Mitosporic Fungi (Deuteromycetes) in the Dictionary of the Fungi. *In: The fungal holomorph* (D.R. Reynolds, J.W. Taylor, eds.). CAB International, Wallingford, 27–55.
Sutton B.C., Hennebert G.L., 1994. Interconnections amongst anamorphs and their possible contribution to ascomycete systematics. *In: Ascomycete systematics, problems and perspectives in the nineties* (D.L. Hawksworth, ed.). Plenum Press, New York, pp. 77–100.
Sutton B.C., Sandhu D.K., 1969. Electron microscopy of conidium development and secession in *Cryptosporiopsis* sp., *Phoma fumosa*, *Melanconium bicolor*, and *M. apiocarpum*. *Can. J. Bot.* **47**, 745–749.
Terracina F.C., 1974. Fine structure of the septum in *Wallemia sebi*. *Can. J. Bot.* **52**, 2587–2590.
Terracina F.C., 1977. Ultrastructure of conidiogenesis in the thallic Hyphomycete *Oidiodendron griseum*. *Can. J. Bot.* **55**, 580–584.
Traquair J.A., Kokko E.G., 1981. Spore discharge in *Epicoccum nigrum*. *Can. J. Bot.* **59**, 59–62.
Tu C.C., Kimbrough J.W., 1973. A rapid staining technique for *Rhizoctonia solani* and related fungi. *Mycologia* **65**, 941–944.
Tubaki K., 1963. Taxonomic study of Hyphomycetes. *Ann. Rep. Inst. for Ferm.*, Osaka I, 25–54.
Tubaki K., 1971. The basauxic conidiophore. Ch. 12. *In: Taxonomy of Fungi imperfecti* (B. Kendrick, ed.). Univ. Toronto Press, Toronto, 176–183.
Vanbreuseghem R., 1966. *Guide pratique de Mycologie médicale et vétérinaire*. Masson, Paris, 206 p.
Vries G.A., de 1952. Contribution to the knowledge of the genus *Cladosporium* Link ex Fries. Thesis, Baarn.
Vuillemin P., 1910a. Matériaux pour une classification rationnelle des Fungi Imperfecti. *C.R. Acad. Sci. Paris*. **150**, 882–884.
Vuillemin P., 1910b. Les Conidiosporés. *Bull. Soc. Sci. Nancy*. **11**, 129–172.
Vuillemin P., 1911. Les Aleuriosporés. *Bull. Soc. Sci. Nancy*. **12**, 151–175.
Wang C.J.K., 1990. Ultrastructure of percurrently proliferating conidiogenous cells and classification. *Stud. Mycol.* **32**, 49–64.

Wang C.J.K., Wilcox H.E., 1985. New species of ectendomycorrhizal and pseudomycorrhizal fungi: *Phialophora finlandia, Chloridium paucisporum* and *Phialocephala fortinii. Mycologia.* **77,** 951–958.

Wang C.J.K., Zabel R.A., 1990. *Identification manual for fungi from utility poles in the Eastern United States.* Allen Press, Inc., Lawrence, Kansas, 356 p.

Wastiaux G., 1994. *La microscopie optique moderne.* Tec & Doc-Lavoisier, Paris, 288 p.

Wilken-Jensen K., Gravesen S. (eds.), 1984. *Atlas of moulds in Europe causing respiratory allergy.* ASK Publishing, Copenhagen, 110 p.

Wingfield M.J., 1985. Reclassification of *Verticicladiella* based on conidial development. *Trans. Br. mycol. Soc.* **85,** 81–93.

Yang C.S., Korf P., 1985. *Ascorhizoctonia* gen. nov. and *Complexipes* emend., two genera for anamorphs of species assigned to *Tricharina* (Discomycetes). *Mycotaxon* **23,** 457–481.

Zheng R., 1985. Genera of the Erysiphaceae. *Mycotaxon* **22,** 209–263.

ABBREVIATIONS
used in the additional references

ABN—*Acta Botanica Neerlandica*, Wageningen.
ADT—*Annales du Tabac. Recherche et développement, section 2*, Paris.
AFSAT—*Annali della Facolta di Scienze Agricoltura, Universita di Torino*, Torino.
Agronomie—*Agronomie*, Paris.
APS—APS Press, St Paul, Minnesota, 1993 (*Ceratocystis and Ophiostoma, Taxonomy, Ecology, and Pathogenicity*, 293 p.).
ARIF—*Annual Report of the Institute for Fermentation*, Osaka.
ASNT—*Annales de la Société des Sciences naturelles de Toulon et du Var*, Toulon.
AVL—*Antonie van Leeuwenhoek*, Wageningen.
Bib. Myc.—*Bibliotheca Mycologica*, J. Cramer, Lehre.
BSMF—*Bulletin de la Société Mycologique de France*, Paris.
CJB—*Canadian Journal of Botany*, Ottawa.
Crypt. Bot.—*Cryptogamic Botany*, Berlin & Stuttgart.
Crypto. Myc.—*Cryptogamie, Mycologie*, Paris.
EJFP—*European Journal of Forest Pathology*, Hamburg & Berlin.
Fitol.—*Fitologiya*, Sofia.
Fl. Jena—*Flora*, Jena.
Fol. G. Pr.—*Folia Geobotanica et Phytotaxonomica*, Praha.
Gams 1971—GAMS W., 1971.—*Cephalosporium-artige Schimmelpilze (Hyphomycetes)*. G. Fischer, Stuttgart, 262 p.
GN 1983—GERLACH W., NIRENBERG H., 1983.—*The genus Fusarium—a pictorial atlas*. Mitt. Biol. Bundesanst. Ld-u, Forstw. Berlin-Dahlem 209, 406 p.
GVR 1982—GVRITISHVILI M.N., 1982.—[*Champignons du genre Cytospora Fr. en URSS*]. Sabchota Sakartvelo Publishing House, 214 p.
Joffe 1986—JOFFE A.Z., 1986.—*Fusarium species: their biology and toxicology*. J. Wiley & Sons, Somerset, New Jersey, 500 p.
KH 1998—KUBICEK C., HARMAN G., 1998. *Trichoderma and Gliocladium*. vol. 1 "Basic Biology, Taxonomy and Genetics", 300 p.
KNAWP—*Koninklijke Nederlandse Akademie van Wetenschappen, Proceedings, Ser. C.*, Amsterdam.
Mik. Fit.—*Mikologiya i Fitopatologiya*, St Petersbourg.
Myc. Pap.—*Mycological Paper*, Kew.
Myc. Res.—*Mycological Research*, Kew.
Mycol.—*Mycologia*, Lancaster.
Mycopath.—*Mycopathologia*, Den Haag.

Mycosyst.—*Mycosystema*, Beijing.
Mycot.—*Mycotaxon*, Ithaca.
N. Hedw.—*Nova Hedwigia*, Berlin & Stuttgart.
NJPP—*Netherlands Journal of Plant Pathology*, Wageningen.
NRK 1975—NAG RAJ T.R., KENDRICK B., 1975.—*A monograph of Chalara and allied genera*. Wilfried Laurier Univ. Press, Waterloo, 200 p.
NZJ Bot.—*New Zealand Journal of Botany*, Wellington.
O.U.P.—Oslo University Press. I. Mat.—Naturv. Klasse N° 22 (1965); N° 24 (1967).
Pers.—*Persoonia*, Leiden.
Pitt 1979—PITT J.I., 1979.—*The genus Penicillium and its teleomorphic states Eupenicillium and Talaromyces*. Academic Press, London, 634 p.
Rev. Myc.—*Revue de Mycologie*, Paris.
RFF—*Revue Forestière Française*, Nancy.
S.A.M.—*Systematic and Applied Microbiology*, Stuttgart & New York.
S.A.N.—*Soobshcheniya Akademii Nauk Gruzinskoï SSR*, Tbilisi.
Stud. Mycol.—*Studies in Mycology*, Baarn.
Syd.—*Sydowia, Annales Mycologici*, Horn.
TBMS—*Transactions of the British mycological society*, London.
T.M.S. Jap.—*Transactions of the mycological Society of Japan*, Tokyo.
TTB 1987—TETEREVNIKOVA-BABAYAN D.N.. 1987.—*Griby roda Septoria v. SSSR;* Yerevan. Akademiya Nauk Armyanskoï SSR, 479 p.
UFA 1988—UECKER F.A., 1988.—*A world list of Phomopsis Names with Notes on Nomenclature, Morphology and Biology*. Mycologia Memoir N° 13, J. Cramer Berlin & Stuttgart, 231 p.
UP 1981—UPADHYAY H.P., 1981.—*A monograph of Ceratocystis and Ceratocystiopsis*. Univ. Georgia Press, Athens, 176 p.
Wind.—*Windahlia*, Göteborg.

GLOSSARY

acervulus: fructification with a concave base, covered with host tissue, enclosed at first, then open to the air following rupture of the subjacent layers (Table I.F e).

Acroblastosporae: group of fungi characterized by the production of acroblastospores (ch. VII).

acroblastospore: conidium produced by holoblastic budding in acropetal chain (Fig. I.21a, ch. VII).

acrogenous: produced at the tip (of the conidiogenous cell or conidiophore) (Fig. I.15a).

acropetal: conidial chain in which the youngest conidium is at the tip (Fig. I.21a).

acrospore: (conidial maturation) in phragmospores, in which the septa are formed successively from the base to the apex (Fig. I.7b).

aleuria, aleuriospore (-conidium): conidium formed by budding and bulging of the tip of the conidiogenous cell, then basal septation. Conidium typically holoblastic, with wide base and often thick walls (Fig. I.24, ch. IV).

Aleuriosporae: group of fungi characterized by the production of aleuria (ch. IV).

allantoid: (conidium) elongated and curved at the ends, like a sausage.

amerosporous: fungus in which the spore is non-septate.

amerospore (-conidium): non-septate conidium (Table I.C).

amphigenous: said of a fruit body that develops on either side of a leaf.

ampulla: apical bulging of conidiophore that produces conidia in the Botryoblastosporae (Fig. I.22), or bears metulae or conidiogenous cells in other groups.

amylolytic: capable of lysing and utilizing starch.

anamorph: asexual mitotic form in the life cycle of a fungus (syn.: asexual, imperfect, conidial stage or form) (Table I.A). Generally characterized by the production of conidia (except in the Agonomycetes, ch. XIII).

anastomose (= clamp connexion): lateral diverticulum of a hypha at a septum, in some Basidiomycetes at the dikaryotic stage (Table I.A).

annellide: enteroblastic conidiogenous cell with percurrent proliferations characterized by close-set rings on the neck, traces of schizolytic secession of successive conidia (Fig. I.27b, ch. XI).

Annellidae: group of fungi producing their conidia on annellides (ch. XI).

Annellophorae: group of fungi producing their conidia on annellophores (ch. V).

annellophore: holoblastic conidiogenous cell with percurrent proliferation, characterized by more or less elongated rings, traces of successive conidial secessions, after retraction of cytoplasm (Fig. I.27a, ch. V).

annellospores: part of Hughes' group III (Table I.D) characterized by conidiogenous cells with percurrent proliferations. Ambiguous term to be avoided on account of the present distinction between Annellophorae (ch. V) and Annellidae (ch. XI). The term 'annellospore' has not been replaced with corresponding terms for the spores of these latter two groups.

anthracnose: any necrotic spot on leaves on which conceptacles appear in the form of brown or black punctuations.

apical growth: most frequent mode of growth of fungal organs, from the apex, at which specialized cytoplasmic structures develop the wall.

appressorium: differentiated extremity of a fungal hypha ensuring its fixation at the site of its penetration of the host.

aristate: having bristles or cilia.

arthric: process of conidiogenesis by fragmentation of a preexisting conidiogenous cell (Fig. I.17, ch. II).

Arthrosporae: group of fungi characterized by the production of arthrospores (ch. II, part 1).

arthrospore (-conidium): conidium resulting from the fragmentation of a preexisting conidiogenous cell (hypha) (Fig. I.17, ch. II).

astomate: (conidioma) lacking an ostiole.

basauxic (conidiophore): growing at the base from a specialized parent cell (Fig. I.28). **(deuteromycete):** having this particular type of growth (ch. XII).

basifugal (conidial chain): acropetal (Fig. I.21a).

blastic: process of formation by budding (Fig. I.20).

Blastosporae *sensu lato*: group of fungi producing their conidia by holoblastic budding: ch. III (Aleuriosporae and Monoblastosporae, ch. IV; Annellophorae and Annelloblastosporae ch. V; Sympoduloasporae, ch. VI; Acroblastosporae, ch. VII; Botryoblastosporae, ch. VIII, Porosporae, ch. IX).

blastospore: in a wider sense, conidium produced by holoblastic budding (Fig. I.20).

blight: a disease characterized by loss of turgidity and collapse of blossoms and/or leaves.

Botryoblastosporae: group of fungi characterized by the production of botryoblastospores (ch. VIII).

botryoblastospore: conidium formed by simultaneous budding at several sporogenous loci on a more or less ampulliform conidiogenous cell (Fig. I.22, ch. VIII).

botryoconidium (-spore): multicellular spore in the shape of a grape bunch (not wall-like dictyospore), example Table IV.C 4.

botryoid (= botryomorph): grouped like grapes.

bristle: a structure that is sterile but often linked to conidiophores and conidiomata, erect, often melanized, and pointed at the tip (Table VI.A9, 12; X.H9).

bulbil: globular, multicellular propagule, lacking differentiated cortex (in the strict sense) (Fig. I.29, ch. XIII).

caespitose: growing in thick bunches (groups of unattached conidiophores, intermediates between free conidiophores, coremium, and sporodochium).

canker: necrosis of cortical tissue developing in depth and not healing completely.

capitulum: fertile head often bulging from a coremium.

caseicolous: growing on cheeses.

caulicolous: localized on the stem.

cellular appendage: appendage (chiefly of a conidium) of cellular origin with a proper wall (Fig. I.11, j-p).

cellulolytic: capable of lysing and utilizing cellulose.

chlamydospore: resting form, very frequent in fungi, constituted of portions of hyphae in which the cytoplasm is condensed and generally surrounded by a thick, melanized wall (Fig. I.16).

cilium: fine, pointed appendage, present in certain conidia (Fig. I.11, a-d).

circinate: more or less rolled at the tip (Tables VI.F2, X.C5).

clypeus: a shield-shaped stroma capping a fruit body (e.g., pycnidium) (Table I.F, a4).

collarette: in phialides, a funnel-shaped parietal formation, extending beyond the meristematic point (Figs. X.3b, X.6).

concolorous: of a single colour, especially in the phragmospores, in which all the cells are of the same colour.

conidial scar: trace of the insertion of the conidium on the conidiogenous cell (Fig. I.13).

conidiogenesis: mode of production of conidia; particularly concerning the relationship between conidiogenous cell and conidium at the level of the wall.

conidioma (-ata): grouping of conidiophores inside a conceptacle or on the surface of a receptacle (Tables I.F).

conidiophore: simple or complex structure located between the vegetative hypha and the conidia, formed by the conidiogenous cell(s), the fertile hypha, and its possible branches (Fig. I.4).

coremium (-a): stipitate conidioma, composed of erect, agglomerated hyphae of non-angular texture (Table I.F g). See also **synnema.**

cortex: peripheral, protective layer or stratum of sclerotia and rhizomorphs (Fig. I.29).

cosmopolitan: growing in many countries, nearly all latitudes, the Old World and the New World and the two hemispheres.

cuneiform: wedge-shaped.

cupule: cup-shaped organ (Table I.F d).

damping-off: collapse and death of seedling plants resulting from the development of a stem lesion at soil level.

deep-seated mycosis: fungal attack on viscera (lungs, liver, kidneys).

deltoid: in the shape of the Greek letter delta, or an equilateral triangle.

dermatomycosis: skin disease caused by a fungus.

dermatophyte: fungus parasite on skin and its productions (e.g., hair, nails).

determinate coremium: coremium in which the fertile zone is limited to the apical part or capitulum (Table I.F g1 and 3).

detriticolous: growing among various debris, often plant debris such as fallen leaves.

dictyospore (-conidium): conidium having longitudinal and transverse septa in the manner of a wall (= muriform conidium) (Table I.C).

didymospore (-conidium): bicellular conidium (Table I.C).

dimidiate: (fruit body) reduced to the upper half. Cf. thyriopycnidium (Table I.F c), and the hemispherical pycnidium of *Leptochlamys* (Table VI.G 16).

dimitic (coremium): having two different types of anatomic texture.

diplothecate: (conidium) having a wall with an inner layer C (Fig. I.9).

discoloured: of various colour, especially in multicellular conidia, in which some cells are more melanized than others.

disjunctor: fertile cell or part of fertile cell in which its death and the rupture of its walls releases the conidium (Fig. II.3).

distoseptate: (phragmoconidium) having an inner layer C forming cells liberated at rupture of the spore wall under pressure (Fig. I.8b, I.9). Cf. **diplothecate.** Ant. **euseptate.**

doliiform: keg-shaped.

dolipore: septal pore with bulging edge, characteristic of the Holobasidiomycetes and certain other Basidiomycetes (Fig. I.3b).

echinulate: wall ornamented with small, acute protuberances resulting from echinulation (Fig. I.10b).

echinulation: detachment of layer A full of pigments (Fig. I.10d).

endogenous: (conidium) produced from the inside of a conidiogenous cell (e.g., Fig. I.26b).

endophialide: phialide surmounted by a long, cylindrical neck. The conidia produced from the venter to the tip move along the neck to be released, e.g., *Chalara* (Fig. X.7, Table X.D1) and similar genera.

endophyte: organism living and proliferating deep inside the tissues of the host plant without causing symptoms.

enteroblastic: (conidium) arising by budding from the inside of the conidiogenous cell, phialide or annellide, and surrounded by the latter's inner layer B1 (Fig. I.6b, I.25, I.26; ch. X, XI).

epiphyte: living on the surface of a plant without being a parasite.

erumpent: organ formed inside a substrate or tissue, becoming more or less external at maturity (Table I.F f).

euseptate: (phragmoconidium) without internal layer C (having hyphal septation) (Fig. I.8a, I.9). Cf. **haplothecate.** Ant. **distoseptate.**

exogenous: (conidium) produced on the exterior of the conidiogenous cell (Figs. I.21, I.25).

fimicolous: growing on dung or manured ground.

foliicolous: growing on living leaves.

fringe: in conidia of rhexolytic separation, remains of the disjunctor or conidiogenous cell at the base of the conidium (Fig. I.14b, b').

fungicolous: growing on other fungi.

fusarioid: (conidium) having pediform base, frequent in the genus *Fusarium* (Table X.B10).

fusiform (fusoid): spindle-shaped, with pointed ends.

geotrichosis: disease due to *Geotrichum*.

germ slit: linear zone in the conidial wall, with least resistance, from which the germ tube extends (Fig. I.12).

germ pore: punctiform region in the conidial wall, of least resistance, from which the germ tube extends (Fig. I.12).

glucophile: (organism) capable of utilizing only simple organic compounds, notably simple sugars and oligosaccharides. Its spores, with strong germinative capacity, and its rapid growth enable it to quickly colonize fresh organic matter. In the fungi these are mainly the Mucorales and also certain Deuteromycetes such as *Penicillium*.

guttulate: (cell) containing droplets that are generally oily.

haplothecate: (conidium) surrounded only by layers A and B (Fig. I.9).

helicosporate: fungus characterized by the production of helicospores.

helicospore: elongated conidium coiled like a spring, on one plane (spiral) or in a cylinder; keg-shaped (helicoidal) (Table I.C).

hilum: scar on the conidium at the point at which it was attached to the mother cell (Fig. I.13).

holoblastic: (conidium) produced by budding, in which the wall is formed from all the parietal layers of the conidiogenous cell (Fig. I.6a, ch. III).

holomorph: term designating a species in all its forms (anamorph(s) and teleomorph).

holospore: (maturation) in a phragmospore, in which the median septum appears first, followed by lateral septa (Fig. I.7a).

holothallic: (conidium) formed from the transformation, by bulging, thickening of the wall, and possible septation, of a conidiogenous cell that has reached its definitive length (Fig. I.19).

hyaline: transparent, glass-like. In present usage, colourless. Ant. **melanized**.

hyperparasite: said of a parasite living on another parasite.

hypha: basic component of filamentous fungi, cylindrical structure a few micrometres in diameter, septate, ramified, containing cytoplasm, and growing apically.

hyphopodium: uni- or bicellular branch, simple or lobate, of a hypha enabling it to anchor itself on the host (Table IV.B 16). Term created by Gaillard in 1891 to designate a structure similar to that of the appressorium. This last term, created by Frank in 1883, should be adopted because of its precedence (cf. *Systema Ascomycetum* 14(1): 29, 1995).

hypocreoid: said of an organ coloured red, yellow, green, blue, violet or dark violet, but not black.

hypophyll: that which develops on the lower surface of leaves.

hypovirulence: in a pathogen, existence of colonies that are weakly pathogenic in themselves, which can immunize the host against attack by virulent colonies.

hysterioid: (fructification) elongated and cleft.

indeterminate coremium: coremium fertile in all its parts (Table I.F g2).

intercalary growth: production of wall by an annular meristematic zone, non-apical.

keratinolytic: capable of lysing and utilizing keratin.

keratinophile: growing on keratin.

keratitis: inflammation of the cornea.

layers of the fungal wall (Figs. I.3, I.8): from the exterior to the interior:
— layer A: granulous, amorphous pellicle, polysaccharidic, and sometimes pigmented.
— layer B: fundamental layer, fibrous, polysaccharidic, possibly divided into melanized outer layer B2 and hyaline inner layer B1.
— layer C: supplementary, internal layer, hyaline, of diplothecate conidia.

lignicolous: growing on wood.

ligninolytic: capable of lysing and utilizing lignin.

lignolytic: capable of lysing and utilizing wood.

macroconidium: in cases in which two types of conidia exist in a Deuteromycete, the larger conidia (often septate) are macroconidia.

macronematous: (conidiophore) differentiated, erect and rather elevated, and thus setting off conidia from vegetative hyphae (Fig. I.4).

medulla: central part of a sclerotium or a rhizomorph (Fig. I.29).

melanization: dark coloration (brown, black) of an organ caused by deposit of melanin in the wall.

melanized: coloured by melanin.

meristem arthrospore: conidium in basipetal chain, resulting from the segmentation of a conidiophore, with or without intercalary growth (Fig. I.18, ch. II, part 2).

meristematic point: location on the conidiogenous cell at which a conidium buds.

meristematic zone: fertile region of a conidiogenous cell in which conidia are produced; zone in which new wall is synthesized. E.g., the region limited by the venter and neck of a phialide.

metula: (on a conidiophore) branch carrying conidiogenous cells (especially in the group *Aspergillus-Penicillium*).

microconidium: the smaller conidium in cases of conidial dimorphism (see **macro-conidium**).

micronematous: (conidiophore) poorly or not differentiated, not carrying conidia at a distance from vegetative mycelium (Fig. I.4a, b).

monilioid, moniliform: (hypha) having swellings at regular intervals, like a pearl string. Cf. *Monilia* (Table VII A 4).

monoblastospore: single holoblastic conidium produced by a more or less differentiated conidiogenous cell (Fig. I.24).

Monoblastosporae: group of fungi characterized by the production of monoblastospores (ch. IV).

monomitic: (coremium) having a single type of anatomic texture.

mucilaginous appendage: mucous appendage, extracellular, without a proper wall (Fig. I.11, e-i).

multicoloured: see **discoloured**.

muriform: (conidium) characterized by longitudinal and transverse septation (like a wall).

muticate: without bristle or cilia. Ant. **aristate**.

mycoparasite: said of a parasite living on a fungus.

obovoid: (shape of an organ: spore, conidioma) ovoid but inverted (the thin end toward the base).

onychomycosis: mycosis of nails.

osmophile: growing well under conditions of high osmotic pressures, as in high sugar concentrations.

ostiole: small opening in the wall of a conidioma (pycnidium, pseudopycnidium), can be located at the tip of a beak or neck.

paraphysis: sterile filament growing along with reproductive cells (Table X.O 3, 4).

parasite: that which feeds on living organic matter.

parent cell: specialized fertile cell producing in basauxic Deuteromycetes, a conidiophore (Fig. I.4).

parenthosome: hemispherical organelle, perforated or not, formed from the endoplasmic reticulum, located on each side of the dolipore in certain Basidiomycetes (Fig. I.3c).

parietal connective: the thin layer A joining into a fragile chain certain arthroconidia that become rounded and tend to separate from each other (Fig. II.3, *Stephanosporium*).

pathogen: organism that causes disease in other living beings.

phialide: conidiogenous cell producing conidia in an enteroblastic way (by budding with continuity between the inner layer of the conidiogenous cell and the conidial wall) without elongation of the phialide (Figs. I.25, I.26, X.1 to 7).

Phialosporae: group of fungi characterized by the production of phialospores (ch. X).

phialospore (-conidium): conidium produced by a phialide.

phragmospore (-conidium): conidium having at least two transverse septa (Table I.C).

plasmodesma (-ata): threads of cytoplasm that pass through multiple pores in the cell walls and join the cytoplasm of adjacent cells (e.g., Fig. II.4).

pleurogenous: formed laterally (on the conidiophore or conidiogenous cell) (Fig. I.15a).

pore: —opening in a hyphal or conidial septum, ensuring communication between cells. Of simple type in the Ascomycetes, in which it can be closed by a Woronin body, and of dolipore type (see **dolipore**) in Holo- and Phragmobasidiomycetes (Fig. I.3).

— conidial scar and hilum, with melanized thickenings in the Porosporae (Fig. I.23, III.19, 20).

Porosporae: group of fungi characterized by the production of porospores (ch. IX).

porospore: conidium characterized by the presence of a pore (sense 2, cf. supra).

proliferation: increase in size of the conidiogenous cell between the production of two conidia. More particularly,

— percurrent proliferation: regrowth across the conidial scar (Figs. I.27; III.11, 12, ch. XI);

— sympodial proliferation: regrowth to the side of the conidial scar (Figs. I.21b; III.3, 6).

pseudopleurogenous: (position of a conidium) initially produced at the tip but pushed out laterally by the proliferation of the conidiogenous cell (e.g., Fig. III.11b).

pseudopycnidium: highly stromatic pycnidium of various forms (Table I.Fb).

pycnidium: closed conceptacle, more or less spherical, having a proper wall and producing conidia (Table I.Fa).

rachis: extended fertile zone of the conidiogenous cell, in the Sympodulosporae. Depending on the modes of sympodial growth, it may have chiefly the shape of a grater (= radula) or a zigzag (sympodula) (Table I.D II).

radicicolous: located at the root of a plant.

rhexolytic: (secession) due to the rupture of the conidiogenous cell wall under the hilum (Fig. I.14b, b').

rhomboidal: (conidium) having the shape of two cones joined at the base (Table VI.A9 to 15).

rhombosporate: fungus characterized by the production of rhomboidal conidia.

root rot: name given to brownish changes in colour seen in culm of cereals close to the soil. The changes result either in dropping of the aerial parts above the injured part (eyespot of cereals) or insufficiency of water supply to the ears (take-all disease).

rostrate: (conidioma) having a beak-shaped appendage (Table I Fa.3).

saprobe, saprophyte: that which feeds on dead organic matter.

schizogenous: (septum) dividing at the zone of cleavage between the two layers B' (Fig. I.14a).

schizolytic: (secession) due to the separation of the parent cell and the conidium at the schizogenous septum (Fig. I.14a).

sclerotium: resting, resistant form of propagule comprising a distinct cortex and medulla (Fig. I.29).

scolecospore: filiform conidium, septate or not, in which the length-width ratio exceeds 15/1 (Table I.C).

scurf: superficial ailment (epidermis, cork) generally due to a multitude of necrotic or suberitic spots.

secession: separation of the conidium and the conidiogenous cell (Fig. I.14).

semimacronematous: (conidiophore) of medium size and differentiation, intermediate between macronematous and micronematous (Fig. I.4c, d).

seminicolous: growing on seeds.

septum: transverse wall of a hypha or conidium (Figs. I.7, I.8, I.9, I.14), generally having:

— a simple pore in the Ascomycetes, which can be closed by a Woronin body (Fig. I.3a)

— a dolipore with parenthesome in the Basidiomycetes (Fig. I.3b).

shot-hole: perforation of leaf lamina from various causes. Holes of fungal origin can be recognized by a small border of scar tissue that crimps the hole after the drop of necrotic tissue caused by the fungus.

sooty mould: black mycelial coating on aerial organs of plants, developing from exudates (= honeydew) of certain stinging insects (aphids, cochineals).

sorus (-i): group of fertile hyphae producing spores.

spiral: rolled in a flat coil (see helicospore).

sporodochium: fertile, pustuliform stroma, superficial or greatly erumpent (Table I.F f).

scab: dark marks spread over plant organs, giving them a multicoloured or speckled appearance, depending on their arrangement and size. On branches they develop into shallow or deep cracks.

staurosporate: fungus characterized by the production of staurospores.

staurospore (-conidium): septate or non-septate conidium, having several arms, like a star, or irregular (Table I.C).

stroma: agglomeration and interlacing of hyphae, often brown, that can hold together fragments of substrate or host tissue, of a determinate or irregular form; it may contain fructifications of the pseudopycnidium type (Table I.Fb).

subcuticular: located between the cuticle and the top layer of epidermal cells.

subiculum: in certain fungi, the modified tissue of the host bearing a fruit body (e.g., pycnidium) (Table I.Fa 5).

sympodial: see proliferation, sympodial.

sympodula: zigzag-shaped tip of a conidiogenous cell, resulting from an alternating sympodial proliferation (Fig. I.21b).

sympodulate: in the shape of a sympodula; characterized by sympodulae.

Sympodulosporae: group of fungi characterized by the production of sympodulospores (ch. VI).

sympodulospore (-conidium): holoblastic conidium produced on a meristematic point resulting from sympodial proliferation (see **proliferation**) of the conidiogenous cell (Fig. III.6; ch. VI).

synanamorph: anamorph linked to one (or several) other(s) in the life cycle of a given fungus (Fig. I.2).

synnema (-ata): bundle of adjacent hyphae. For fertile structures of this type see **coremium**. (The term synnema is now used in English in the same sense for fertile structures.)

taxon: a taxonomic group of organisms occupying the same rank in the classification (e.g., genus, species).

teleomorph: meiotic, sexual morph characterized by the production of asci-ascospores, basidia-basidiospores (Fig. I.1; Table I.A). Syn. sexual or perfect stage or form.

telluric: growing in the soil.

thallospore (-conidium): conidium arising from a preexisting piece of thallus (or from a fertile hypha, or part of a conidiophore, Fig. I.17); as opposed to blastospore, a true conidium; without growth after initiation of conidial primordium (Fig. I.16, 19).

thyriopycnidium: flat, superficial conidioma, in the form of a dimidiate shield with radiate structure with or without central foot (Table I.F c).

tretic: (conidiogenesis) characteristic of porospores.

tretoconidium: see porospore.

turbinate: in the shape of a top or club (Table VI.A 11, 12).

ubiquitous: in a large number of countries, widespread (but less so than 'cosmopolitan').

Uredinales: or 'rusts', one of two orders of the class Teliomycetes characterized by septate basidia. The rusts are obligatory parasites on many plants.

verrucose: ornamented with small warts resulting from a detachment of the parietal layer A, larger than in echinulations and more or less empty (Fig. I.10c, e).

wall: the layer, external to the cytoplasm, surrounding the fungal organs, hyphae, and spores (Fig. I.3).

Woronin body: hexagonal or spherical organelle that, in the Ascomycetes, serves to close the septal pore when needed (Fig. I.3a).

INDEX OF TAXA

(The names of teleomorphs are underlined. Names of taxa other than genera and species are in italics.)

The numbers in bold type indicate the pages on which the taxon is drawn and the other numbers pages on which the taxon is cited.

For Product Safety Concerns and Information please contact our EU
representative GPSR@taylorandfrancis.com
Taylor & Francis Verlag GmbH, Kaufingerstraße 24, 80331 München, Germany

www.ingramcontent.com/pod-product-compliance
Ingram Content Group UK Ltd.
Pitfield, Milton Keynes, MK11 3LW, UK
UKHW051941210425
457613UK00026BA/66